WILDERNESS OF WILDLIFE TOURISM

Advances in Hospitality and Tourism

WILDERNESS OF WILDLIFE TOURISM

Edited by

Johra Kayeser Fatima, PhD

Apple Academic Press Inc.
3333 Mistwell Crescent
Oakville, ON L6L 0A2 Canada

Apple Academic Press Inc.
9 Spinnaker Way
Waretown, NJ 08758 USA

First issued in paperback 2021

ISBN 13: 978-1-77-463691-6 (pbk)
ISBN 13: 978-1-77-188481-5 (hbk)

Library and Archives Canada Cataloguing in Publication

Wilderness of wildlife tourism / edited by Johra Kayeser Fatima, PhD.

(Advances in hospitality and tourism book series)
Includes bibliographical references and index.
Issued in print and electronic formats.
ISBN 978-1-77188-481-5 (hardcover).--ISBN 978-1-315-36581-7 (PDF)
1. Wildlife-related recreation. 2. Tourism. I. Fatima,
Johra Kayeser, author, editor II. Series: Advances in hospitality
and tourism book series
SK655.W55 2017 338.4'791 C2016-908268-7 C2016-908269-5

Library of Congress Cataloging-in-Publication Data

Names: Fatima, Johra Kayeser, editor.
Title: Wilderness of wildlife tourism / editor, Johra Kayeser Fatima.
Description: Oakville, ON ; Waretown, NJ : Apple Academic Press, 2017. |
Includes bibliographical references and index.
Identifiers: LCCN 2016057231 (print) | LCCN 2016059575 (ebook) | ISBN 9781771884815 (hardcover : alk. paper) | ISBN 9781315365817 (ebook)
Subjects: LCSH: Wildlife watching industry. | Wildlife-related recreation. | Ecotourism.
Classification: LCC G156.5.E26 W537 2017 (print) | LCC G156.5.E26 (ebook) | DDC 338.4/759--dc23
LC record available at https://lccn.loc.gov/2016057231

Apple Academic Press also publishes its books in a variety of electronic formats. Some content that appears in print may not be available in electronic format. For information about Apple Academic Press products, visit our website at **www.appleacademicpress.com** and the CRC Press website at **www.crcpress.com**

ABOUT THE EDITOR

Johra Kayeser Fatima, PhD

Dr. Johra Kayeser Fatima is an Assistant Professor in the School of Management, University of Canberra, Australia. She received her PhD from the University of New South Wales, Sydney, Australia. She has more than 10 years of experience teaching business and tourism courses both nationally and internationally, such as in Australia and Bangladesh.

Dr. Fatima has published in different prestigious journals, including *Tourism Analysis, Environmental Education Research, International Journal of Quality and Reliability Management, Asia Pacific Journal of Marketing and Logistics*, and *Social Responsibility Journal*. She regularly speaks at academic conferences around the world and shares her views on various research themes such as sustainable tourism, tourism for economic development, relationship marketing, and services marketing. As a key research team member, she also worked in various research institutes in the past, such as the *Centre for Sustainable and Responsible Organizations*, Deakin University, Australia; and the *Centre for Economic*, and *Health Convergence*, McGill University, Canada. She currently holds a membership in the *American Association for Women in Higher Education*, USA. She can be reached at johra.fatima@canberra.edu.au.

**Dedicated with love to my lion cub,
Rafan Khan**

CONTENTS

LIST OF CONTRIBUTORS

Ruchi Badola
Wildlife Institute of India, Dehradun, Uttarakhand, India

Debasish Batabyal
Pailan School of International Studies, Kolkata, West Bengal, India

Jeremy Buultjens
School of Business and Tourism, Southern Cross University, Lismore, New South Wales, Australia. E-mail: Jeremy.buultjens@scu.edu.au

Aparup Chowdhury
Bangladesh Parjatan Corporation, Bangladesh

Li Cong
Tourism Management Department, Landscape School, Beijing Forestry University, Beijing, China. E-mail: congli1980@163.com

Aditi Dev
Wildlife Institute of India, Chandrabani, Dehradun, Uttarakhand, India

Pariva Dobriyal
Wildlife Institute of India, Chandrabani, Dehradun, Uttarakhand, India

Johra K. Fatima
School of Management, University of Canberra, Australia. E-mail: johra.fatima@canberra.edu.au

V.S. Avila-Foucat
Instituto de Investigaciones Económicas, Universidad Nacional Autónoma de Mexico, Mexico. E-mail: savila@iiec.unam.mx

Amanat Kaur Gill
Wildlife Institute of India, Chandrabani, Dehradun, Uttarakhand, India

Athula Chammika Gnanapala
Department of Tourism Management, Sabaragamuwa University of Sri Lanka, Belihuloya, Sri Lanka

Azizul Hassan
Cardiff Metropolitan University, Cardiff, Wales, UK. E-mail: M.Hassan15@outlook.cardiffmet.ac.uk

Md. Ziaul Haque Howlader
Bangladesh Parjatan Corporation, Bangladesh. E-mail: ziabpc@gmail.com

Syed Ainul Hussain
Wildlife Institute of India, Chandrabani, Dehradun, Uttarakhand, India

A. Aguilar Ibarra
Instituto de Investigaciones Económicas, Universidad Nacional Autónoma de Mexico, Mexico

Mahmood A. Khan
Virginia Polytechnic and State University, Blacksburg, VA, USA

Natalie King
International Tourism and Hospitality Management, Taylor's University, Subang Jaya, Malaysia. E-mail: natoking@yahoo.com

Diane Lee
Tourism and Events Program, School of Arts, Murdoch University, Perth, Western Australia

Thounaojam Sanggai Leima
Wildlife Institute of India, Chandrabani, Dehradun. Uttarakhand, India

Vikneswaran Nair
Sustainable Tourism Management, Taylor's University, Subang Jaya, Malaysia

David Newsome
Environment and Conservation Sciences Group, School of Veterinary and Life Sciences, Murdoch University, Perth, Western Australia

Md. Golam Rabbi
Wildlife & Nature Conservation Circle, Forest Department, Bangladesh. E-mail: rabbi_rk@yahoo.com

Iraj Ratnayake
Department of Tourism Management, Sabaragamuwa University of Sri Lanka, Belihuloya, Sri Lanka

Nilanjan Ray
Netaji Mahavidyalaya, Kalipur, West Bengal, India. E-mail: nilanjan.nray@gmail.com

Anukrati Sharma
Department of Commerce and Management, University of Kota, Rajasthan, India. E-mail: dr.anukratisharma@gmail.com

Bitapi C. Sinha
Protected Area Network, Wildlife Management and Conservation Education Department, Wildlife Institute of India, Dehradun, Uttarakhand, India. E-mail: bcs_wii@yahoo.co.uk

A. Sanchez-Vargas
Instituto de Investigaciones Económicas, Universidad Nacional Autónoma de Mexico, Mexico

Sneha Thapliyal
Wildlife Institute of India, Chandrabani, Dehradun, Uttarakhand, India

R. M. Wasantha Rathnayake
Department of Tourism Management, Sabaragamuwa University of Sri Lanka, Belihuloya, Sri Lanka. E-mail: warath1@gmail.com

Bihu Wu
International Centre for Recreation and Tourism Research, School of Urban and Environmental Sciences, Peking University, Beijing, China

LIST OF ABBREVIATIONS

BSMSP	Bangabandhu Sheikh Mujib Safari Park
CCB	community capacity building
CCTV	closed circuit television
CIP	consumer involvement profile
CNP	Corbett National Park
CPW	collaborative partnership on sustainable wildlife management
CS	consumer surplus
CV	contingent valuation
CVM	contingent valuation method
CWT	consumptive wildlife tourism
DDC	Dolphin Discovery Centre
DoF	Department of Forest
DPI	deep place involvement
DWLC	Department of Wildlife Conservation
ECA	ecologically critical area
FAM	familiarization
FD	Forest Department
GPS	global positioning system
GR	game reserve
HH	household
HVC	high conservation value
IUCN	International Union for Conservation of Nature
KLNP	Keibul Lamjao National Park
MESCOT	model ecologically sustainable community conservation and tourism
MPI	medium place involvement
NCWT	non-consumptive wildlife tourism
NDBR	Nanda Devi Biosphere Reserve
NDNP	Nanda Devi National Park
NGOs	non-governmental organizations
NOAA	National Oceanic and Atmospheric Administration
NP	National Park

NSP	Nishorgo Support Project
NTCA	National Tiger Conservation Authority
NTO	National Tourism Organization
OLS	ordinary least square
PA	protected areas
PF	protected forest
PII	personal involvement inventory
RF	reserved forest
SANDEE	South Asian Network for Development and Environmental Economics
SNP	Satchari National Park
SPI	shallow place involvement
TCM	travel cost method
TPB	theory of planned behavior
UNDP	United Nations Development Program
USP	unique selling preposition
VAB	value-attitude-behavior hierarchy
VOF	Valley of Flowers National Park
VR	visitation rate
WII	Wildlife Institute of India
WOM	word of mouth
WS	wildlife sanctuaries
WTA	willingness to accept
WTP	willingness to pay
WVO	wildlife value orientation
WWF	World Wildlife Fund for Nature

ADVANCES IN HOSPITALITY AND TOURISM BOOK SERIES BY APPLE ACADEMIC PRESS, INC.

Editor-in-Chief:
Mahmood A. Khan, PhD
Professor, Department of Hospitality and Tourism Management,
Pamplin College of Business,
Virginia Polytechnic Institute and State University,
Falls Church, Virginia, USA
Email: mahmood@vt.edu

Books in the Series:

Food Safety: Researching the Hazard in Hazardous Foods
Editors: Barbara Almanza, PhD, RD, and Richard Ghiselli, PhD

Strategic Winery Tourism and Management: Building Competitive Winery Tourism and Winery Management Strategy
Editor: Kyuho Lee, PhD

Sustainability, Social Responsibility and Innovations in the Hospitality Industry
Editor: H. G. Parsa, PhD
Consulting Editor: Vivaja "Vi" Naraparеddy, PhD
Associate Editors: SooCheong (Shawn) Jang, PhD,
Marival Segarra-Oña, PhD, and Rachel J. C. Chen, PhD, CHE

Managing Sustainability in the Hospitality and Tourism Industry: Paradigms and Directions for the Future
Editor: Vinnie Jauhari, PhD

Management Science in Hospitality and Tourism: Theory, Practice, and Applications
Editors: Muzaffer Uysal, PhD, Zvi Schwartz, PhD, and
Ercan Sirakaya-Turk, PhD

Tourism in Central Asia: Issues and Challenges
Editors: Kemal Kantarci, PhD, Muzaffer Uysal, PhD, and Vincent Magnini, PhD

Poverty Alleviation through Tourism Development: A Comprehensive and Integrated Approach
Robertico Croes, PhD, and Manuel Rivera, PhD

Chinese Outbound Tourism 2.0
Editor: Xiang (Robert) Li, PhD

Hospitality Marketing and Consumer Behavior: Creating Memorable Experiences
Editor: Vinnie Jauhari, PhD

Women and Travel: Historical and Contemporary Perspectives
Editors: Catheryn Khoo-Lattimore, PhD, and Erica Wilson, PhD

Wilderness of Wildlife Tourism
Editor: Johra Kayeser Fatima, PhD

Medical Tourism and Wellness: Hospitality Bridging Healthcare (H2H)©
Editor: Frederick J. DeMicco, PhD, RD

Sustainable Viticulture: The Vines and Wines of Burgundy
Claude Chapuis

The Indian Hospitality Industry: Dynamics and Future Trends
Editors: Sandeep Munjal and Sudhanshu Bhushan

ABOUT THE SERIES EDITOR

Mahmood A. Khan, PhD, is a Professor in the Department of Hospitality and Tourism Management, Pamplin College of Business at Virginia Tech's National Capital Region campus. He has served in teaching, research, and administrative positions for the past 35 years, working at major U.S. universities. Dr. Khan is the author of seven books and has traveled extensively for teaching and consulting on management issues and franchising. He has been invited by national and international corporations to serve as a speaker, keynote speaker, and seminar presenter on different topics related to franchising and services management. He is the author of *Restaurant Franchising: Concepts, Regulations, and Practices, Third Edition, Revised and Updated*, published by Apple Academic Press, Inc.

Dr. Khan has received the Steven Fletcher Award for his outstanding contribution to hospitality education and research. He is also a recipient of the John Wiley & Sons Award for lifetime contribution to outstanding research and scholarship; the Donald K. Tressler Award for scholarship; and the Cesar Ritz Award for scholarly contribution. He also received the Outstanding Doctoral Faculty Award from Pamplin College of Business.

He has served on the Board of Governors of the Educational Foundation of the International Franchise Association, on the Board of Directors of the Virginia Hospitality and Tourism Association, as a Trustee of the International College of Hospitality Management, and as a Trustee on the Foundation of the Hospitality Sales and Marketing Association's International Association. He is also a member of several professional associations.

PREFACE

The world is a wide surface for love, laughter, and learning. Learning may begin with assisted lessons in formal educational institutions or may be self-generated by observing nature. While the wild nature initially seems to be full of mess, a deep understanding unveils the truth ... every single movement in nature is following a rule ... a rule of life. Thus, a great enthusiasm has been found in today's tourists to get closer to the wildlife in nature to entertain themselves, to fulfill learning and ... to see the rule of life.

Life varies in every corner of this whole natural world. So, this book is a little effort to grab the essence of wildlife tourism in different parts of the world as depicted by local tourism researchers in every chapter. Since the never-ending exploration of wildlife tourism persists, this thirst of knowing the unknown motivates us to keep the title of the book as "Wilderness of Wildlife Tourism."

—Johra K. Fatima, PhD

CHAPTER 1

A BRIEF CONSIDERATION OF THE NATURE OF WILDLIFE TOURISM

DAVID NEWSOME*

Environment and Conservation Sciences Group, School of Veterinary and Life Sciences, Murdoch University, Perth, Western Australia, Australia

**E-mail: D.newsome@murdoch.edu.au*

CONTENTS

ABSTRACT

Wildlife tourism is very broad in its scope and requires the attention of many disciplines in order to understand the specific organism, whole ecology, and social dimensions of the subject. Whale shark tours in Australia provide a good example of sustainable wildlife tourism.

> **"...wildlife tourism is a complex mix of the social, biological and ecological sciences"**

Today wildlife tourism is a complex mix of the social, biological, and ecological sciences. Understanding its human and ecological dimensions involves the protected area management aspects of environmental science, wildlife behavioral ecology, physiology, recreation ecology, biochemistry, statistics, human psychology, tourism studies, economics, marketing, environmental policy, and legislative frameworks. This mix of eclectic subjects makes its study complex and multidisciplinary in nature and is additionally complicated because of the involvement of a wide range of species, locations, and management scenarios. Species differ greatly in their tolerance to human disturbance as well as sites requiring different management protocols and infrastructure depending on the tourism product. Furthermore, there are different modes of access (on foot, via car, and via boat), and humans differ greatly in their attitudes and expectations. People may want to photograph, touch, and feed wildlife, and human understanding of wildlife differs according to age, sex, and culture. Moreover, there are many different types wildlife tourism products around the world (e.g., see Table 1.1). These products range from bird watching, feeding birds, viewing coral reefs, night walks, and the viewing of nocturnal species to swimming with whales, dolphins, and whale sharks.

The success of whale shark tours in Western Australia has led the industry being considered as an example of world best practice (Mau, 2008). One of the best ways to assess quality is via the client education, sustainable practices, and training processes of tour operators and the experiences of tourists engaged in whale shark activities. Patterson (2008) explored tourist satisfaction, perceptions of potential negative impacts of the tours on the sharks, compliance with management rules, whether there was a high level of tour operator management to reduce potential impacts, and whether there was any education and research conducted into whale shark conservation. Patterson (2008) confirmed that the industry operates according to strict regulations and a code of conduct such as adhering to a maximum ratio of swimmers to sharks at any one time of 10:1, swimmers to approach to only a maximum of 3 m from the shark, and no touching permitted. Motorized floatation devices and the use of "duck-diving" technique or the use of SCUBA gear are also not permitted.

In terms of the boat-based aspects of the operation, boats are to approach no closer than 30 m from the sharks with a maximum speed of 8 knots. No other vessel is allowed to come within 250 m of contact zone of the boat that is currently engaged in whale shark interactions. Additional protection is afforded in that vessels may spend a maximum of 60 min within the 250 m contact zone of the shark during an interaction.

TABLE 1.1 Wildlife Tourism Products—An Australian Perspective.

Wildlife tourism product	Example	Location
Glow-worm viewing	Visit to glow-worm sites	Lamington National Park, Qld
Viewing coral reef biota	Snorkel, dive and glass bottom boat tours	Ningaloo Reef, WA
Fish feeding	Glass bottom boat tours	Ningaloo Reef, WA
Swim with whale sharks	Boat-based snorkel tours	Ningaloo Reef, WA
Swim with manta rays	Boat-based snorkel and dive tours	Ningaloo Reef, WA
Cage diving with great white sharks	Boat-based snorkel and dive tours	Neptune Islands, SA
Viewing and feeding crocodiles	Boat-based tours	Adelaide River, NT
Bird watching	Tour operator led bird watching tours	All states, e.g. Coates Wildlife Tours, WA
Bird feeding	Parrot, whipbird and honeyeater feeding	O'Reilly's Lamington National Park, Qld
Nocturnal mammal viewing in the wild	Tour operator led night spotting tours	Atherton Tableland Qld
Nocturnal mammal viewing within a fenced facility	Private reserves National Park Conservation Centers	Karakamia Sanctuary, WA Barna Mia, Dryandra National Park, WA
Nocturnal mammal feeding	Tasmanian Devil Restaurants	Cradle Mountain, Tas

TABLE 1.1 *(Continued)*

Wildlife tourism product	Example	Location
Dolphin feeding	Beach interaction zone	Monkey Mia, WA
Swim with dolphin tours	Boat-based snorkel and dive tours	Bunbury, WA
Whale watching	Boat-based tours	Hervey Bay, Qld
Swim with whales	Boat-based tours	Minke Whale Tours, Qld
Visits to seal and sealion haul out-sites and breeding colonies	Boat-based tours	Penguin Island and Jurian Bay, WA

Note: WA (Western Australia), Tas (Tasmania), SA (South Australia), NT (Northern Territory), and Qld (Queensland).

Patterson (2008) conducted surveys with tourist participants and found that the level of control, monitoring of codes of conduct, education, and conservation responsibility to be of a high standard in comparison with other existing marine wildlife tours, including other whale shark tours in the North-West and overseas in the Philippines. The series of monitoring techniques, tour practices, and research and education policies with whale sharks were recognized as a benchmark for emerging wildlife tours around the world.

Given the high standards of environmental management evident in Western Australia, while at the same time recognizing potential problems with emerging marine wildlife tourism operations elsewhere in the world, Patterson's work inspired the development of a marine wildlife tourism assessment framework. The framework, as a contribution to fostering sustainable wildlife tourism, consists of assessing the ecological, environmental, social, and operational aspects of marine wildlife tourism (see Rodger et al., 2011 for details). When tested under field conditions, Rodger et al. (2011) concluded that it was useful for identifying and collating existing information of tourism interest and was able to identify gaps in knowledge and areas of concern. I thus emphasize that the challenge for future is to foster, practice, and maintain quality. One way of doing this is to increase our understanding of how we can access and view nature without disturbance. We also need to be mindful of how we interact with wildlife, especially with regard to feeding practices, taking photographs, and the human desire to touch and sometimes manipulate a wild animal (see Newsome et al., 2005).

KEYWORDS

- **wildlife tourism**
- **whale shark tours**
- **sustainable practices**
- **Australian perspective**

REFERENCES

Mau, R. Managing for Conservation and Recreation: The Ningaloo Whale Shark Experience. *J. Ecotour*. **2008,** *7* (2–3), 213–225.

Newsome, D.; Dowling, R.; Moore, S. *Wildlife Tourism;* Channel View Publications: Clevedon, UK, 2005.

Patterson, P. An Examination of Ecotourism with Whale Sharks (*Rhincodon typus*) at Coral Bay: Assessment of Vital Social Stakeholders; M.Sc. Dissertation, Department of Environmental Science, Murdoch University, Australia, 2008.

Rodger, K.; Smith, A.; Newsome, D.; Moore, S. Developing and Testing a Rapid Assessment Framework to Guide the Sustainability of the Marine Wildlife Tourism Industry. *J. Ecotour.* **2011,** *10,* 149–164.

CHAPTER 2

PLANNING AND DEVELOPMENT OF WILDLIFE TOURISM IN BANGLADESH

MD. ZIAUL HAQUE HOWLADER* and APARUP CHOWDHURY

Bangladesh Parjatan Corporation, Dhaka, Bangladesh

**Corresponding author. E-mail: ziabpc@gmail.com*

CONTENTS

ABSTRACT

In the context of present population size of Bangladesh, wildlife conservation is a very challenging issue for the country. On the other hand, development of tourism industry for employment generation, community benefits and poverty reduction is also now a government priority sector. The Tourism Policy, 2010 prioritized sustainable tourism development based on wildlife and nature. Many challenges are taking place in the country for wildlife conservation and tourism development. The government of Bangladesh emphasizes eco-tourism development at some protected areas (PAs) of the country. Bangladesh being the largest deltaic plain along with tropical climate, its wildlife is diverse and unique in nature. The country is proud of possessing many unique wildlife animals. The world's biggest cat, *Panthera Tigris Tigris*—royal Bengal tiger is a wonder of nature. The bio-diversity of the Sundarbans also attracts many wildlife lovers every year. To lure both the domestic and foreign tourists, Bangladesh has many flagship species, mainly in the PAs, including the tiger, Asian elephant, hoolock gibbon, and olive reef turtle. Bangladesh is moving ahead in an effort to promote wildlife tourism by offering package tours in organized manner. Although Bangladesh is yet to pull in many a number of tourists, at present it has good wildlife management policy, law and program, wildlife conservation and management projects as well as tourism development policy and program. Wildlife tourism is getting popular because of increasing awareness of local people, and love to nature of tourists. The Wildlife Act, 2012 of Bangladesh defines about the sanctuary, ecopark, ecotourism, national park (NP), biological diversity, forest products, wild animals, protected wildlife, and so forth. Article 18(a) of Constitution of Bangladesh reads that biodiversity conservation is a constitutional obligation to all citizens of the country.

> **"Bangladesh is moving ahead in an effort to promote wildlife tourism..."**

Bangladesh being the largest deltaic plain along with tropical climate, its wildlife is diverse and very unique in nature. The country is proud of possessing many unique wildlife animals that raise the eyebrow of wildlife experts as well as tourists. The world's biggest cat, *Panthera Tigris Tigris*—Royal Bengal Tiger is a wonder of nature. The bio-diversity of the Sundarbans also attracts many wildlife lovers every year.

According to the Forest Department of Bangladesh, "the country is a transitional zone of flora and fauna, because of its geographical settings and climatic characteristics. There are many rivers and streams in the country covering a total length of 22,155 km. About 11% of the country's area belongs to different types of water bodies. In addition to the regular inland waters, seasonally a large part of the country remains submerged for 3-4 months during monsoon. Haor basin in north-east region of the country is such an important wetland. The wetland system is a vast repository of bio-diversity." Tangua Haor—a Ramsar Site is rich in biodiversity. In winter millions of migratory birds fly in this haor. It is a bird sanctuary.

Bangladesh is abode of about 53 species of amphibian, 19 species of marine reptiles, 158 species of reptiles (including 19 marine species), 690 species of birds (380 resident, 209 winter visitors, 11 summer visitor, and 88 vagrant), 121 species of mammals and five species of marine mammals.[1] In addition to the large bird count, a further 310 species of migratory birds swell bird numbers each year. Every year new species are also added in the national checklist.

To lure both the domestic and foreign tourists, Bangladesh has many flagship species, mainly in the protected areas (PAs), including the tiger, Asian elephant and hoolock gibbon, and olive reef turtle (Batagur baska). Each of these flagship species plays a key role in its respective ecosystems and is necessary for ensuring the delicate ecological balance. The tiger is the national animal of Bangladesh. Many globally threatened birds can be easily viewed in Bangladesh, including notably the white-rumped vulture, Pallas's fish eagle, lesser adjutant, and masked finfoot. Although the country has no endemic species of wildlife, the Bostami soft-shell turtle was treated as endemic until in 2007 when it was established that the species also occurs in Assam, India. Although not globally threatened, the estuarine crocodile is threatened nationally and it is treated as a flagship species for the aquatic ecosystem of the Sundarbans. Other than the flagship or threatened species there are many interesting minor species.

Bangladesh is moving ahead in an effort to promote wildlife tourism by offering package tours in organized manner. Though Bangladesh is yet to pull in many a number of tourists like African destinations, at present it has good wildlife management policy, law and programs, wildlife conservation and management projects as well as tourism development policy

[1]Department of Forest, GoB

and programs. Wildlife tourism is getting popular because of increasing awareness of local people, and love to nature of domestic and international tourists. Bangladeshi tourists also tend to visit the natural attractions as well as the wildlife animals.

The Wildlife Act, 2012 of Bangladesh has clearly defined about the sanctuary, ecopark, ecotourism, botanical garden, national park (NP), biological diversity, sacred tree, forest products, wild animals, protected wildlife, core zone, buffer zone, and so forth. This act also adds clauses for stringent punishment against any poaching, hunting, and killing of any wild animal including migratory birds.

Article 18(a) of Constitution of Bangladesh reads that biodiversity conservation is a constitutional obligation to all citizens of the country.

The National Tourism Policy, 2010 emphasizes the planning and development of wildlife tourism through proper management. The Bangladesh Parjatan Corporation Ordinance, 1972 also delineates the development of wildlife tourism in different forms.

In Bangladesh, the private tour operators conduct package tours by following the forest rules and regulations. Tour operators brief tourists about the dos and don'ts prior to entering into the forest. Recently, the government has developed "Sundarbans Travel Rules" which has clearly described the role of tour operators, tourists, and forest department. These rules are very much conducive to wildlife protection in the Sundarbans as well as wildlife tourism development.

To promote wildlife tourism as well as proper conservation of wildlife, government has declared protected forest (PF), reserved forest (RF), NP, eco-park, safari park, botanical garden, and so forth, also co-management through local community participation.

The Government has until now established about 17 NP, 21 sanctuaries, two botanical gardens, two safari parks, and eight ecoparks [2] among which to be mentioned are—Bhawal National Park in Gazipur, Boleswar DC Park in Pirojpur, Madhupur National Park in Tangail and Mymensingh, Ramsagar National Park in Dinajpur, Himchari National Park in Cox's Bazar, and Lawachara National Park in Moulavi Bazar. Apart from these, Bangladesh is the repository of the wonder of the world, the national pride—the Sundarbans—abode of the majestic Royal Bengal tiger.

[2]Department of Forest http://www.bforest.gov.bd/site/page/7304f3af-8d7b-4fcd-a237-41b5be4de286;

Bhawal National Park in Gazipur, 40 km north of Dhaka, the capital of Bangladesh with 5022 ha area, of which 2000 ha are used for recreation. The forest is mainly dominated by the *Sal* (Shorea robusta) trees. The *Sal* trees are a very captivating species because of their being indigenous to this area as well as in other areas of Bangladesh. This NP possesses 6 mammals, 10 amphibians, 9 reptiles, and 39 birds[3]. There are foot trails, fishing area, Northern River Terrapin turtle breeding and conservation site, zoo, butterfly gardens which attracts a large number of eco-tourists. The government enforces a very good forest management plan here.

The *Altadighi* is another interesting place for promoting tourism as well as preservation and conservation of wildlife and biodiversity. This NP is located in Naogaon district. This park covers 264 ha which is dominated by a large pond 1.2 km long and 200 m wide, some small patches of Sal forest, golden jackals and over 60 bird species. Every year huge number of migratory birds also visits the park. This NP is also attractive to many national and foreign tourists because of its diversified wildlife.

Madhupur National Park is at a distance of about 80 miles distance from Dhaka that takes about 1 h to reach. The prominent wildlife to be mentioned are leopard, wild hog, wild cow, spotted deer, wild cock, peacock, jackal, wild cat, mongoose, red mouth monkey, wild goat, black mouth baboon, porcupine, squirrel, hawk, vulture, wild buffalo, mynah, kite, nightingale, swallow, pigeon, dove and kingfisher, skylark, sparrow, owl, woodpecker, parakeet, hare, pangolin, parrot, martin, and so forth.

Located about 5 km south of Cox's Bazar town, Himchhari National Park was established in 1980 which was once a passage for the Asian Elephants. It is now a place for tourists to escape Cox's Bazar town to see patches of evergreen forest. There is a waterfall there, which is best seen in the rainy season. Its area is 1729 ha. The park comprises the reserve forest areas of Bhangamera and Chainda blocks under Cox's Bazar Forest Department. Evergreen and semi-evergreen tropical forests are found in this area. There are about 55 species of mammal, 286 bird species, 13 amphibians, 56 reptile species, and 117 plant species. The other outstanding features are 286 species of birds, occasional elephants sighting, beautiful waterfall, and so forth.

Lawachara National Park located 60 km south of the Sylhet city in the Komalganj Upajila of Moulvibazar District is one of those few PAs

---

[3]Bangladesh Forest Department Report, 2015

in Bangladesh where tourists can watch hoolock gibbons. It is basically a plantation raised during 1920's for timber production by conversion for years; The plantation has taken a structure very similar to natural forests and supports wildlife of different kinds. The park is important in regulating water flows, serves as a watershed forming important catchments. Ecologically this semi-evergreen hill forest is a transition between the Indian-subcontinent and the Indo-China floristic region. The main wild-lives are slow loris, pig-tailed macaque, rhesus macaque, Assamese macaque, capped langur, phayre's leaf-monkey, hoolock gibbon, jackal, wild dog, sloth bear or Himalayan black bear, yellow-throated marten, tiger, leopard, fishing cat, leopard cat, wild pig, sambar, barking deer, and Indian giant squirrel.

The Sundarbans—the UNESCO world heritage site. This mangrove of serene natural beauty with peace and tranquility is one of the world heritage sites in the country. It is a cluster of islands densely forested and crisscrossed by hundreds of meandering streams, creeks, rivers, and estuaries. It is one of the richest repositories of biodiversity with 330 species of mangrove plants, and 424 species of wildlife including 315 species of birds. It is a natural habitat of spotted dear, crocodiles, wild boar, lizards, monkeys, pythons, king cobras and of course kingly Royal Bengal Tiger as per the latest census by camera-trapping method. This heavily forested swampy land is the last strong hold of more than 106 royal Bengal tigers. The life style of the wood-cutters, nomadic fishermen and the daring honey collectors have enhanced the charm of this unique beauty. Following are the most famous wildlife and tourists' attractions within this forest:

1. *Hiron* Point (*Nilkamal*) for tiger, monkey, deer, birds, crocodiles, and natural beauty.
2. Katka for deer, tiger, varieties of birds and monkey, crocodiles, and morning and evening symphony of wild fowls. A vast expanse of grassy meadows running from *Katka* to *Kachikhali* (tiger point) provides opportunities for wild tracking.
3. *Tin Kona* Island is a good place for having glance of tiger and deer.
4. Dublar Char (Island) for fishermen. It is a beautiful island where herds of spotted deer are often seen grazing.
5. *Karamjal* Mangrove Arboretum for viewing dolphin jumping, and deer and crocodile rearing station.

6. *Kachikhali* for virgin beach night canoeing deer, crocodiles, monkeys, and monitor lizards.
7. *Mandarbaria* for turtle breeding spot, dolphins, king crabs, and numerous species of crabs.

The Swatch of No Ground and Marine PA located to the south of the Bay of Bengal near the the Sundarbans which is about 1738 km^2 is a home for different kind of dolphins (Iraboti, Gangetic, spotted, etc.), turtles, sharks, and so forth. This is the paradise for wildlife-loving tourists. Visiting the Swatch of No Ground tops the itineraries of the Sundarbans going tourists.

Department of Forest (DoF) under the Ministry of Environment and Forests of Bangladesh administers 1.53 million ha of forest land mainly under the legal categories of RF and PF. PAs in Bangladesh have been declared from the eighties decade under the Wildlife Preservation (Amendment) Act, 1974. This is an important criterion followed for designating existing RFs and PFs under PAs related to conserving biological diversity in all the bio-geographical zones of Bangladesh.

The basic principle of PA management is that every PA develops a management plan that guides and controls the management of the resources within the PA, the conservation of biodiversity, the uses of area and the development of PA facilities. A majority of the country's PAs were gazetted by covering existing RFs and PFs that were hitherto managed by following the prescriptions made under the respective working plans. Separate management plans were prepared in 1997 for implementing specific prescriptions as per the established principles of PA management.

The Government of Bangladesh has established PAs in different forest types of all the four identified bio-ecological zones (tropical evergreen and semi-evergreen forests, moist deciduous forests, mangrove forests, and reed land and wetland forests). Presently there are 16 notified PAs under the management of Bangladesh Forest Department, which covers 241,675 ha lands under three PA categories—NP, wildlife sanctuaries (WS), and game reserve (GR). Most of these PAs were initially managed as per each one of the working plans under "Preservation Working Circle." After enacting the Wildlife Preservation (Amendment) Act, 1974, Wildlife Circle was created in 1976 with specific responsibility for wildlife policy related matters.

The government has also implemented many wildlife management and tourism development projects by the assistance of World Wide Fund

(WWF), Nature Conservation, World Bank, Asian Development Bank (ADB), and so forth. At present the Department of Forest is implementing a World Bank—sponsored project named "Strengthening Regional Cooperation for Wildlife Protection." This project would provide regional support through capacity building of the relevant agencies in each country (Bangladesh, Bhutan, and Nepal) to address the illegal trafficking, trading, and poaching of wildlife commodities.

The International Union for the Conservation Of Nature (IUCN) Bangladesh Chapter is extensively engaged for protecting the wildlives in the country. This organization has undertaken various kinds of programs in this regard, amongst which to be mentioned are "Updating Red Species Lit of Bangladesh," "Environment Management and Biodiversity Conservation Plan for Sundarbans' Biodiversity," "White-rumped Vultures Conservation in Bangladesh," "Eco-system for Life", and so forth.

There are strong co-management plans and programs for the PAs in Bangladesh. Local people are actively engaged in the co-management programs. Local people promote tourism by providing services to tourists like lodging (home stay), foods and guide services. Local people are directly and indirectly benefited owing to the co-management and wildlife tourism development. Bangladesh Parjatan (Tourism) Corporation has also recently entered into agreement with the two organizations (*Ajier* and CBT Bangladesh) for community-based tourism development in Bangladesh. These two organizations also promote community-based tourism around the PAs. Bangladesh Parjatan Corporation also provides training to local youth on tour guiding. These tour guides can work anywhere in Bangladesh including the wildlife areas.

Every year, the number of nature lovers and wildlife watchers are increasing as the ticket selling of national parks, botanical garden, zoo, and PAs indicates. The revenue generations from these sites are also on the rise. Private tour operators of Bangladesh also are offering package tours.

Bangladesh being a young destination in world tourism map, she is trying to diversify her tourism products and make a sustainable tourism development. Among the many other programs of the government for tourism development, and wildlife tourism would be a very good initiative. Hence, there should be a coordinated effort of the government bodies especially between the Ministry of Civil Aviation & Tourism and Ministry of Environment & Forest. Bangladesh Parjatan (tourism) Corporation and

Bangladesh Tourism Board may go for a special kind of marketing initiative on wildlife tourism.

KEYWORDS

- **wildlife management**
- **tourism development**
- **employment generation**
- **community benefit**

REFERENCES

Altadighi Ecotourism Management Plan 2015–2025, Department of Forest: Dhaka, Bangladesh, April 2015.

Annual Report, BFD (2014–2015).

Bahwal Ecotourism Management Plan 2015–2025, Department of Forest, Dhaka, Bangladesh, April 2015.

Khan, M. Monirul H. *Protected Areas of Bangladesh: A Guide to Wildlife.* Nishorgo Program, Wildlife Management and Nature Conservation Circle, Bangladesh Forest Dept.: Dhaka, Bangladesh, 2008.

National Tourism Policy, 2010; Government of Bangladesh.

Wildlife Act- 2012; Government of Bangladesh.

CHAPTER 3

ASSESSMENT OF RECREATIONAL SERVICES OF NATURAL LANDSCAPES IN THIRD WORLD TROPICS USING THE TRAVEL COST METHOD

RUCHI BADOLA, SYED AINUL HUSSAIN*, PARIVA DOBRIYAL, THOUNAOJAM SANGGAI LEIMA, AMANAT KAUR GILL, ADITI DEV, and SNEHA THAPLIYAL

Wildlife Institute of India, Chandrabani, Dehradun, Uttarakahand, India

Indian Institute of Management Indore Prabandh Shikhar, Rau-Pithampur Road Indore - 453556, Madhya Pradesh, India", E-mail: snehat@iimidr.ac.in.

Corresponding author. E-mail: hussain@wii.gov.in

CONTENTS

ABSTRACT

Travel cost method (TCM) is a widely used tool for the evaluation of the recreational value of natural landscapes with potential for promoting nature-based tourism. Consumers' surplus can be used to increase the cost of visit for maintaining high-value low-volume tourism, especially in the fragile and culturally and ecologically vulnerable areas. This chapter presents the recreational value of three biodiversity rich and culturally significant landscape of India. In this study, we used TCM to derive the economic value of the nature-based tourism at Nanda Devi Biosphere Reserve (NDBR), Corbett National Park (CNP) and Keibul Lamjao National Park (KLNP). Individual approach was used to collect the information at both KLNP and CNP while zonal approach was used at NDBR. Personal interviews with tourists were conducted at all the selected sites. NDBR, CNP, and KLNP generate huge recreational value but most of the revenue generated either goes to the strongest players in the area or outsiders who can afford to provide better facilities to the visitors. The leakage of the tourism benefits due to forgone consumers' surplus, leads to the loss of benefit accrued by the local communities resulting inadequate economic capital to invest in the tourism industry. Nature-based recreation and tourism is publicized as a sustainable way to preserve nature while providing economic benefits to the local communities. These additional benefits provide opportunity for generating alternative livelihood for local communities and invest in improved basic facilities to promote sustainable tourism.

> **"Wildlife tourism has provided alternative livelihood to the people in the area...."**

3.1 INTRODUCTION

The sustainable development goal recommends policies that promote sustainable tourism in order to create new livelihood opportunities for the local communities and to support their traditional income generation strategies. Tourism has emerged as a mechanism to generate employment, combat poverty, and promote sustainable development. It provides benefits that represent a significant part of the total economic value of nature in modern societies, and are an increasingly important determinant in multi-functional forest management. Tourism promotes national amalgamation, understanding between nations and supports local economy. Overexploitation and increased social cost of natural resources is an important issue on the global agenda. Environmental valuation methods have been applied to value and calculate these benefits and losses to sustainably manage the natural resources (Barbier et al., 1997; Gürlük & Rehber, 2008).

Recreation is one of the copious services provided by ecosystems. The users attach a value to nature recreation, which is substantial and not reflected by market prices as it is provided as a quasi-public good (Zandersen & Tol, 2009). On a practical level, considering these values can result in better management, conservation, and planning activities for nature recreation. On a research level, advancing knowledge on the range of values of ecosystems dependent population characteristics, quality and quantity of the natural resource as well as specification of demand models, is essential when assessing general trends and impacts on the use of forests for recreation. The recognition of the economic benefits of recreation has provided a sound economic rationale globally and has been identified as a tool for biological conservation (Ahmed et al., 2007).

Natural areas are frequently the focus of recreational trips but seldom command a price in the market. Tourism in mountain ecosystems is increasing at a rapid pace for scenic beauty, culture, history, and adventure-recreational opportunities that mountains offer. This growth provides benefits to local communities and national economies. On the other hand, if unmanaged and developed in unsustainable way it poses a potential threat to the ecological health and culture heritage (UNEP-CI, 2007) and inequitable benefit sharing.

The value of a service may change due to change in climate, land use conversion, and other natural and anthropogenic causes. It is necessary to allocate and value these services to ensure their continued supply in the future and to develop effective policies and raise funds to generate alternative sustainable livelihood for local communities and manage ecological health of the area. Understanding the potential value of forest recreation and other non-marketed benefits and quantification of such values is not new (Willis & Benson, 1989; Chaudhry & Tewari, 2006) but serious attempts to implement the finding of these studies are comparatively few.

This study aims to evaluate the recreational service provided by three protected landscapes of the India. Valuation of recreational use of the area will help in management and development of a sustainable ecotourism model for equity in benefit sharing in biodiversity hot spot and ecologically fragile areas.

3.2 STUDY SITES

3.2.1 NANDA DEVI BIOSPHERE RESERVE (NDBR)

The great wilderness of the Nanda Devi region in the Indian Himalayas has been recognized as a Biosphere Reserve and a World Heritage Site for its unique natural diversity (Bosak, 2008). NDBR (30° 05′–31° 02′ N Latitude, 79° 12′–80° 19′ E Longitude) is located in the state of Uttarakhand, a Himalayan state in India, with an area of 6020.43 km^2 (Fig. 3.1). The NDBR comprises of two core zones (Nanda Devi National Park (NDNP), 630 km^2; Valley of Flowers National Park (VOF), 87.5 km^2) surrounded by a buffer zone and a transition zone. Both the core zones have been declared as world heritage sites by UNESCO. No human habitation is present inside the core zones but there are 47 villages in the buffer zone and 33 villages are in the transition zone. Bhotia (Indo-Mongoloid) and Garhwali (Indo-Aryan) are the main communities in the area. Communities are entirely dependent on natural resources for cultural, agricultural, and other livelihood activities (Silori, 2004). Nanda Devi peak, the second highest peak of India, located inside the NDNP, was once one of the most popular destinations for mountaineering and expedition till the ban on tourism in 1982, which has affected the local economy and has resulted in conflict between the management authorities and local communities (Silori, 2004).

3.2.2 KEIBUL LAMJAO NATIONAL PARK (KLNP)

KLNP is the only natural home of the rare and endangered Eld's deer (*Rucervus eldii eldii*) locally known as "*Sangai.*" The Park lies between latitude 24° 26′ N and 24° 31′ N and longitude 93° 49′ E and 93° 52′ E in the south-eastern fringes of the Loktak Lake Manipur, India, occupying an area of 40 km^2 (Fig. 3.1). The unique feature of the Park is the floating meadows, locally known as "*phumdi,*" which is formed by the accumulation of organic debris and biomass with soil particles. The Park received national and international attention when Loktak Lake was declared as a site of international importance on March 23, 1990, under the Ramsar Convention. Despite its protected status, KLNP is under enormous anthropogenic pressure and in the absence of alternative livelihood opportunities local communities are dependent on the resources provided by the Park. People visit park for both its cultural and natural value.

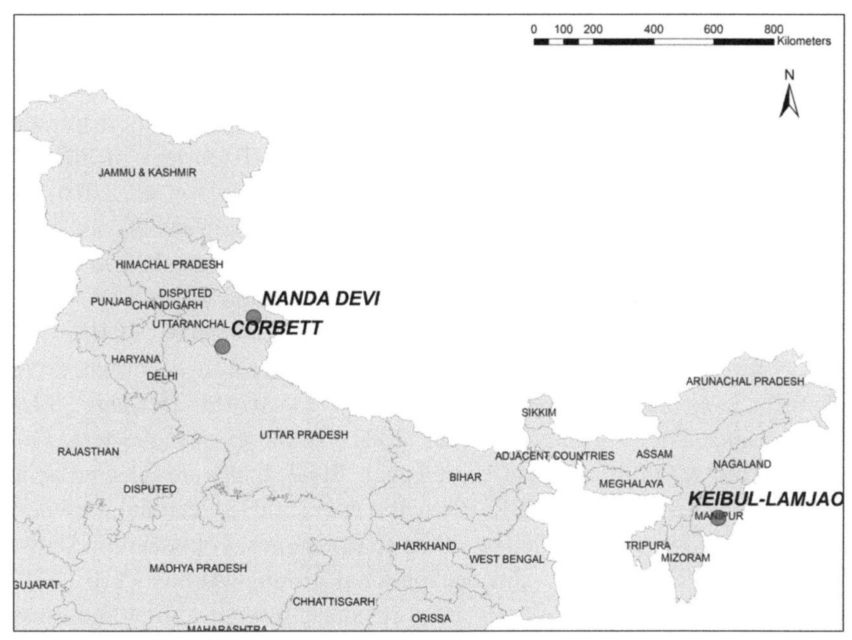

FIGURE 3.1 Map showing locations of all three study sites (Nanda Devi Biosphere Reserve, Corbett National Park and Keibul Lamjao National Park), India.

3.2.3 CORBETT NATIONAL PARK (CNP)

The CNP is situated in the ecologically important Bhabar-Tarai region, a strip of land skirting the southern part of the Shivalik Mountains. It is located in the Kumaon and Pauri-Garhwal region of Uttarakhand state. The geographical extent of the area is from 78° 05′ E to 79° 08′ E longitude and 29° 25′ N to 29° 48′ N latitude (Fig. 3.1). The park is home to rich and diverse faunal and floral species. Though there are no human settlements inside the national park, it is surrounded by 92 villages whose residents depend on the buffer zone forests for fuel wood, fodder, and grazing livestock. After the declaration of the national park as a core zone of the tiger reserve, some villages were left just near the fringes of the core zone with no buffer in between. The buffer forests also serve as habitat for spill-over population of wildlife from the core zone, leading to incidents of human wildlife conflict. Wildlife tourism has provided alternative livelihood to the people in the area.

3.3 METHODOLOGY

The travel cost method (TCM) was used to collect the data for estimating the economic value of wildlife and nature-based tourism in all three selected sites (Clawson, 1959; Moeltner, 2003; Badola et al., 2010). We used the zonal approach in NDBR as most of the tourists do not repeat their trip to the area due to its remoteness (Guha & Ghosh, 2009). Individual approach was used to collect the information at both KLNP and CNP, since both the sites get repeated visitors (Badola et al., 2010).

A semi-structured questionnaire was developed in accordance with UNESCO's manual "Managing Tourism at World Heritage Sites" (Pedersen, 2002; Kuosmanen et al., 2004). A survey was pre-tested and carried out at sampling sites (Ahmed et al., 2007) that were selected on the basis of the visitation rate (VR, Petrosillo et al., 2007). Personal interviews with tourists were conducted at all the selected sites. Tourists representing different socio-economic and occupational groups (Akhter et al., 2009) and with different profile (e.g., local resident, foreign tourist and tourist on independent holiday, or on package tour) and purpose of visit at a particular site were interviewed (Badola et al., 2010). For zonal approach, the sampled population was divided into zones on the basis of the place of origin of the tourists. Areas of origin of tourist were not divided into circular zones as getting population estimation for these zones was difficult (English & Bowker, 1996); for this study, we used states as the zones.

While CNP is open to visitors/tourists only during winter and summer; KLNP and NDBR are visited by tourists throughout the year with maximum visitations recorded during winter and summer. Sampling in KLNP and NDBR was done throughout the year, that is, during both high- and low-visitation seasons to avoid the overestimation of the recreational value. In CNP, the sampling was done when the park is open for tourists. To normalize information on the different variables utilized in the study, distance was measured as the distance between the capital of the state of origin and the sampling site (in kilometers). For tourist profile, information about income, age, education status, and state of origin was collected. To calculate the travel cost information about the mode and cost of travel, information on group composition, cost of stay and food, and expenditure on services provided by local people was gathered. Time spent on travel (to and fro) and time of stay in the area was evaluated on the basis of hourly wages (Mcconnell & Strand, 1981; Guha & Ghosh, 2009).

The total cost of travel is calculated for each visit by summing the total cost of travel, the monetary value of the time spent in travel to get to site and at the site and money spent at the site for stay, food, hiring of guides, and procurement of local products. The cost of time spent is estimated using the wages per hour for a particular visitor. Cost of travel to the site was considered for round trip and was calculated by summing cost of distance travelled, and fare paid for journey and maintenance cost for vehicle used, which was calculated per km per visit. Maintenance cost included cost of fuel, money spent to refurbish, or other maintenance work of the vehicle. These clubbed together gives the total cost of travel. The regression analysis between visits and travel cost was done to get an equation that relates visits per capita to travel costs (Badola et al., 2010). A total of 781 interviews with visitors were conducted at all three sites of which 361 were at NDBR, 112 at KLNP, and 308 at CNP (Table 3.1).

TABLE 3.1 Tourism Profile of Nanda Devi Biosphere Reserve (NDBR), Keibul Lamjao National Park (KLNP) and Corbett National Park (CNP), India.

Place	Type of tourism	No. interview conducted	Season of maximum visits	Season for visit
NDBR	Nature based and cultural	361	Summer	Throughout the year
KLNP	Nature based and cultural	112	Winters	Throughout the year
CNP	Nature based and wildlife	308	Winter and Summer	Winter and Summer

Individuals above the age of 18 years were approached and on confirming their willingness to participate in the questionnaire survey were interviewed. Initially 841 tourists were approached, of which 60 tourists refused to grant the interview; hence, the response rate was about 93%. Respondents were not offered any monetary or non-monetary benefits.

3.4 RESULTS

3.4.1 *RECREATIONAL VALUE OF THE NDBR*

3.4.1.1 TOURIST PROFILE

Of the 361 tourists sampled, 79.8% were male and only 20.2% were female; 85% people said that they were visiting the place for the first time while 15% had visited the area earlier. Among the repeat visitors, 55.6% were visiting the place for the second time while 18.5% had more than 10 visits in previous years. Mean age of respondents was 36.8 ± 0.6 (SE). Group size of the tourist was ranged between 22 and single person (mean = 6.36, SE = 0.31). Respondents were categorized into different educational, occupational, and income groups and classes. Out of the respondents, 48.5% were graduates followed by postgraduates (34.6%). Least percent (1.4%) of respondents were Ph.D. Respondents belonged to eight different occupational backgrounds. More than half of the respondents were working in the private sector (53.7%), 17.7% were self-employed and had their own business, 12.5% were working with different government agencies and 8.6% were students. The respondents were sorted into seven different income classes (I–VII). Majority of the respondents belonged to income class III (US$ 323–615.39/month) followed by the class I (not earning, which included elderly, housewives, and students), IV (US$ 630.77–923.08/month), and VI (US$ 1246.15–1538.46/month). Least number of respondents belonged to the income class II (US$ 15.39–307.69/month) (Table 3.2).

3.4.1.2 DEMAND CURVE AND CONSUMERS' SURPLUS FOR NDBR

VR for each zone was calculated by dividing estimated visits from the zone by its population. Curve estimation was used and non-linear inverse curve fit obtained through the equation VR = −13.015 + ((3.67 × 10^5)/ travel cost). The plot of the fitted model of the demand curve generated through regression represents the whole experience demand curve. Various models were applied on the data and were tested for their predictive ability (based on R^2); the inverse model (V/P = a + b/C) of the fitting was finally selected, as the coefficient of determination was found to be highest (R^2 = 0.688) for this model. From the plot of the curve, it is observed that VR becomes almost constant at travel cost US$ 369.23. The above whole-experience demand curve has been used for creating a "net recreational demand curve" by adding a hypothetical increment in the travel costs like US$ 15.39, 30.77, 46.15, and so forth, and forecasting the number of tourists. To estimate the consumers' surplus, a regression was again carried out with VR as the dependent variable and added travel costs as the independent variable. The curve was fitted through non-linear curve estimation and the logarithmic model ($VR = 4681.7e^{-2E\text{-}04\,(\text{travel cost})}$) was found to be more appropriate ($R^2 = 0.998$) to fit the curve (Fig. 3.2). Area under this curve gives the consumers' surplus per year, that is, US$ 14,828 for recreational use of the NDBR. Dividing this figure by average number of visit ($n = 22{,}048$ per year) gave the consumers' surplus per visits per year (US$ 0.67).

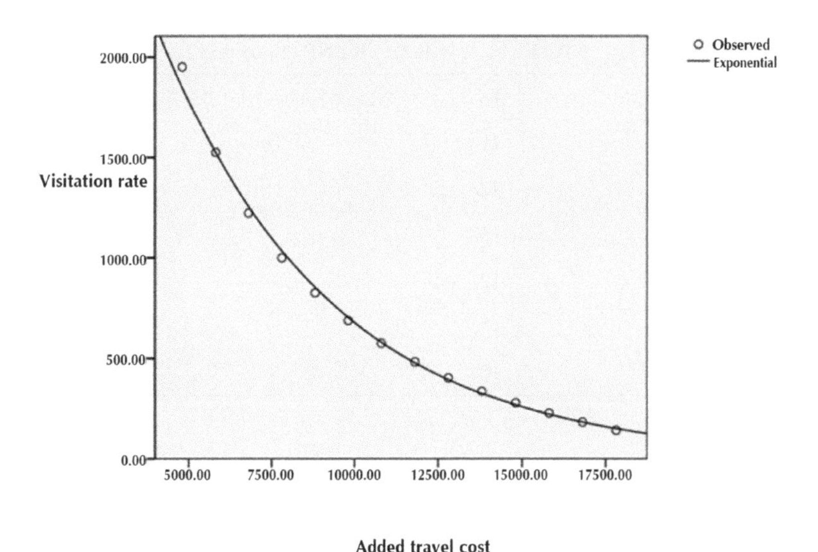

FIGURE 3.2 Net recreational demand curve for recreational use of Nanda Devi Biosphere Reserve.

3.4.2 *RECREATION VALUE OF KLNP*

3.4.2.1 *PROFILE OF TOURISTS*

Of the 112 tourists interviewed 81% were male and only 19% were female. Most of the visitors (44%) were between the age group of 18–30 years followed by 31–40 years (24%). About 39% of the tourists were graduates and 15% had completed post graduation. None of the visitors were illiterate and had at least some years of formal education. While 14 and 23% were unemployed and students, respectively, the rest of the visitors were engaged in government jobs (24%), private jobs (21%), or were self-employed (18%). Most of the visitors (43%) fall in the less than US$ 46.15 per month income group because majority of the respondents were students or unemployed. Visitors in the US$ 76.92–153.85 per month income group constituted 22% of the total visitors while people with more than US$ 384.62 monthly income composed only 5% of tourists (Table 3.2).

TABLE 3.2 Income Classes of Tourists Visiting Nanda Devi Biosphere Reserve (NDBR), Keibul Lamjao National Park (KLNP) and Corbett National Park (CNP), India.

Income class (US$)	NDBR (%, *n* = 361)	KLNP (%, *n* = 112)	CNP (%, *n* = 308)
Not earning	16	–	–
15.4–307.7	11	79	13
323.08–615.4	22	18	70
630.8–923.1	15	5	17
938.5–1230.8	12	–	–
1246.1–1538.5	14	–	–
More than 1538.5	10	–	–

(US$1 = INR 65)

3.4.2.2 DEMAND CURVE AND CONSUMERS' SURPLUS FOR KLNP

The function obtained for the demand curve a power regression model was fitted (on the basis of $R^2 = 0.998$) and $y = 1830.4x^{-0.087}$ equation was used. The value of the ticket fee was replaced with successively higher values to get the consumers' surplus. In Figure 3.3, the first point on the curve was the total number of visitors at the current access cost (in this case no entry fee). The subsequent points were calculated by estimating the number of visitors with different hypothetical entrance fees until the number of visitors becomes zero and travel cost demand curve for the site was obtained.

Finally, the recreational value was estimated by calculating the area under the demand curve, that is, the consumers' surplus and was worked out to be US$ 0.29 per visit, and when extrapolated to the total number of visitors the total economic benefit from recreation use was estimated to be US$ 1657.69 per annum.

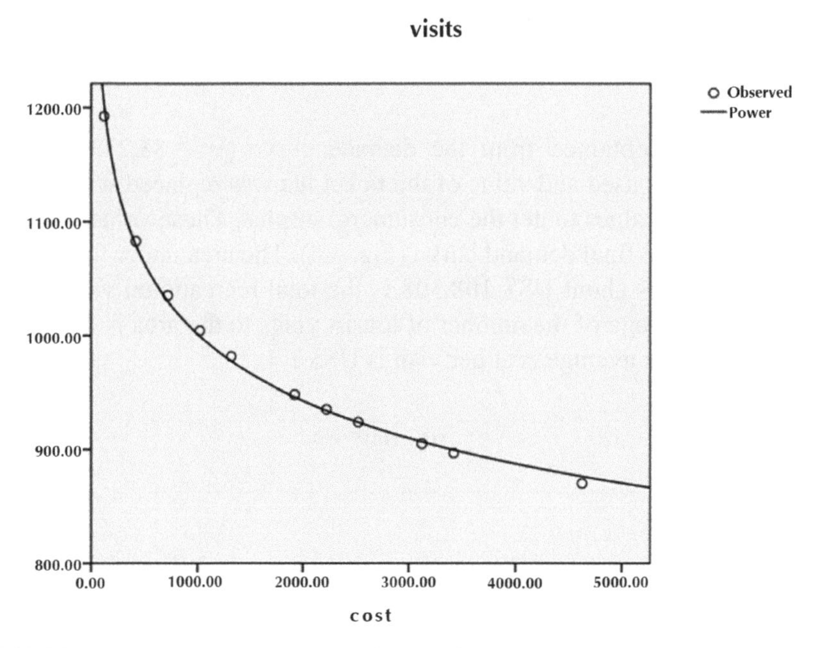

FIGURE 3.3 Net recreational demand curve for recreational use of Keibul Lamjao National Park, India.

3.4.3 RECREATION VALUE OF CNP

3.4.3.1 PROFILE OF TOURISTS

CNP is a major tourist destination for national and international tourists. As the cost of travel and cost of stay is high in CNP, the tourists visiting the CNP generally belong to a higher income group as compared to other sites (Table 3.2). Average group size visiting the area is six. The average number of days spent at CNP is 2–3 days. Most of the visitors belonged to income category US$ 323.08–615.4 (70%) followed by income category US$ 630.8–923.1 (17%) while least number of visitors belonged to income category US$ 15.4–307.7 (13%). The Tigers were found to be the main crowd puller. Though most of the tourists were on leisure trip, a substantial number was also nature lovers.

3.4.3.2 DEMAND CURVE AND CONSUMERS' SURPLUS FOR CNP

The function obtained from the demand curve (y = 88,243–551.98x; R^2 = 0.99) was used and value of the ticket fee was replaced with successively higher values to get the consumers' surplus. These values are used to calculate the final demand curve (Fig. 3.4). The area under the demand curve, which is about US$ 108,308 is the total recreational value of the CNP. The average of the number of tourist visits to the area is 77,612 per year; hence the average cost per visit is US$ 1.4.

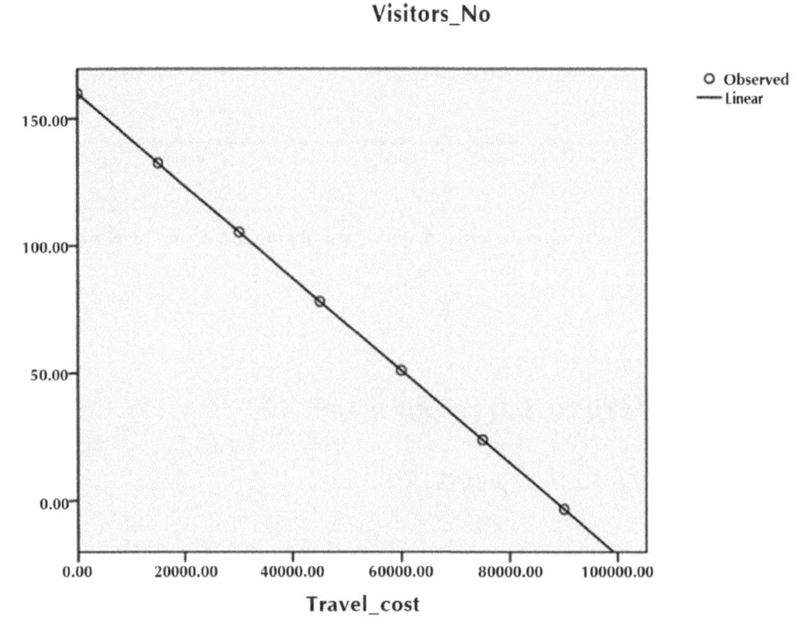

FIGURE 3.4 Net recreational demand curve for recreational use of Corbett National Park, India.

3.5 DISCUSSION

The United Nations World Tourism Organization declared poverty reduction as one of its primary themes for tourism development. Sustainable Tourism-Eliminating Poverty initiative (http://www.unwto.org/step/) focuses on supporting the goals of sustainable tourism activities to alleviate poverty. Lack of economic data on the usage of parks has resulted in an information blind spot due to which natural areas, which provide numerous economic and social benefits, have been assigned a zero price (Union, 1998; Prideaux & Falco-Mammone, 2007). This has resulted in undervaluation of many natural sites as their true economic significance has not been estimated ultimately leading to their degradation.

Two alternative methodologies exist to estimate the recreational value, the contingent valuation method (CVM) and TCM. CVM asks potential visitors what they are willing to pay to visit a recreational site. However, this method does not record visitors' actual behavior (Guha & Ghosh, 2009). In developing countries, CVM cannot always correctly evaluate the recreational value of an environmental resource because of a parallel economy involving different categories of middle to upper income groups of visitors (Chaudhry & Tewari, 2006). Existing literature shows that when there is a market for the service to be valued, CVM should be avoided (Guha & Ghosh, 2009). TCM has been used as a tool to calculate the recreational value of protected and non-protected areas in both developed and developing countries. It provides the exact amount of the consumers' surplus, which can be implemented to increase the cost of visitation mainly by increasing entry fee, travel cost, or price of local products. Increased monetary benefits can be used for generating livelihood opportunities for local communities, which can encourage them to support and participate in the conservation and management of natural resources and wildlife. The additional income in terms of consumers' surplus can also be used to promote and provide improved basic amenities in the area to facilitate the high-end value tourist and ensure sustainable tourism activities in the area.

Result of TCM has been used at few sites and entry fee at some of the sites has even been revised (Mishra, 2014). Strategies should be developed to consider the results of TCM to realize the actual recreational value of a natural site and to ensure maximum tourism benefits to local communities. Though, TCM provides better information for management of a natural recreational site in the developing country scenario, there are certain factors which need to be dealt carefully with, such as (a) measurement of critical variables like price and quantity (Ward & Loomis, 1986; Das, 2013) and (b) estimation of a more inelastic demand curve than the true demand curve, when the travel cost to a given site and the travel cost to a substitute site are positively correlated (Caulkins et al., 1986). The utility function should be consistent with the recreationists' decision-making behavior and the resultant ordinary demand function with underlying preferences (Das, 2013).

Nature-based recreation and tourism are publicized as a sustainable way to preserve nature while providing economic benefits to the local communities (Gössling, 1999; Wunder, 2000; Wood, 1984). Though the mainstay of the tourism industry in India continues to be domestic travelers, tourism is emerging as the second largest foreign exchange earner for the country (Government of India, 2002). The barriers to improved performance of the tourism industry in India include distance of tourist destinations from affluent tourist markets, lack of facilities, relatively limited professionalism in the industry and the "image" of the country (e.g., it is not a holiday location, it poses safety concerns, and it has inadequate services) (Government of India, 2002). The emphasis of the central government on other development issues has affected the plan for tourism with little central help or coordination in many potential regions to develop facilities and promote the available recreational services. Visitor satisfaction is a particularly important consideration in protected area tourism as satisfaction scores are used as a measure of success, or failure, of a protected area to deliver a high-quality visitor experience. Examples where the discourse of valuing protected areas for their economic benefits via tourism (and thus the importance of satisfactory visitor experiences) are frequently noted in public policy statements and documents released to the public.

The NDBR, CNP, and KLNP have large recreational value but most of the revenue generated either goes to the strongest stakeholders of the area or rich outsiders who can invest and can provide better facilities to the visitors. The leakage of the tourism benefits leads to the loss of benefit accrued by the local communities, which lack the economic capital to invest in the tourism industry. NDBR receives a large number of nature-based tourists every year but do not have enough basic facilities, such as health, road, and sanitation for the visitors. Larger benefits go to the rich and strong stakeholders and outsiders while local people only get marginalized benefits. The leaked monetary benefits can be used to combat these issues and to develop and maintain these basic facilities in the area and promote sustainable nature-based tourism in the ecologically fragile areas. Similarly, in CNP most of the tourism benefits are accrued by the economically better off section of the society and economic stratification has taken place. KLNP is the home of endangered Eld's deer (*Rucervus eldii eldii*), which is endemic to the area and have rare floating meadows but, due to the remoteness of the area and lack of basic amenities gets very low number of visitors which ultimately result in high dependence on natural resources and poor economic condition of the local communities. Promotion of the sustainable wildlife tourism can be a tool to address these issues.

The abstained benefits can also be used to compensate the communities, which do not generally get benefits of the conservation of the wild habitat and tourism attracted by such areas but often pay the cost of such conservation program in terms of loss due to human wildlife conflict. As the tourism industry will develop in a sustainable way, it will not only reduce the direct dependence on forests by providing alternative livelihood opportunities but will also lead to improved basic infrastructure in the area. The power of the local community to attain resources (alternative resources from market) will improve and lead to self-sustenance (Lindberg, 2003; Scheyvens, 1999). Being employed in tourism sector provides a regular source of income to many households, and thus economic security. It helps people meet basic financial needs, as well as acquire additional assets or livestock investments to support existing livelihood strategies of local people and ultimately reduces the direct dependence on natural resources. Future research should validate the results of TCM in developing countries and undeveloped societies and also develop strategies to implement the results of the TCM on the ground to improve the quality of local communities and natural resource conservation and management.

3.6 ACKNOWLEDGMENTS

This study was carried out under the project "An integrated approach to reduce the vulnerability of local community to environmental degradation in the Western Himalayas" and "Ecology of Sangai, Manipur, India" funded through the Grant-in-Aid funds of the Wildlife Institute of India

(WII). We thank the Director and the Dean, WII for logistic and technical support. We would like to thank Prof. Mahmood Khan, Head, Department of Hotel, Restaurant and Institutional Management at Virginia Polytechnic Institute, and State University, Prof. Johra Fatima for their help, support, and encouragement.

KEYWORDS

- **travel cost method**
- **recreational value**
- **consumers' surplus**
- **Nanda Devi Biosphere Reserve**
- **Keibul Lamjao National Park**
- **Corbett National Park**

REFERENCES

Ahmed, M.; Umali, G. M.; Chong, C. K.; Rull, M. F.; Garcia, M. C. Valuing Recreational and Conservation Benefits of Coral Reefs- the Case of Bolinao, Philippines. *Ocean Coastal Manage.* **2007,** *50* (1), 103–118.

Akhter, S.; Rana, M. P.; Sohel, M. S. I. Protected Area an Efficacy for Ecotourism Development: A Visitors' Valuation from Satchari National Park, Bangladesh. *Tigerpaper.* **2009,** *36* (3), 1–7.

Badola, R.; Hussain, S. A.; Mishra, B. K.; Konthoujam, B.; Thapliyal, S.; Dhakate, P. M. An Assessment of Ecosystem Services of Corbett Tiger Reserve, India. *Environmentalist.* **2010,** *30* (4), 320–329.

Barbier, E. B.; Acreman, M.; Knowler, D. *Economic Valuation of Wetlands: A Guide for Policy Makers and Planners;* Ramsar Convention Bureau: Gland, Switzerland, 1997; p 138.

Bosak, K. Nature, Conflict and Biodiversity Conservation in the Nanda Devi Biosphere Reserve. *Conser. Soc.* **2008,** *6* (3), 211–224.

Caulkins, P. P.; Bishop, R. C.; Bouwes, Sr. N. W. The Travel Cost Model for Lake Recreation: A Comparison of Two Methods for Incorporating Site Quality and Substitution Effects. *Am. J. Agric. Econ.* **1986,** *68* (2), 291–297.

Chaudhry, P.; Tewari, V. P. A Comparison between TCM and CVM in Assessing the Recreational Use Value of Urban Forestry. *Int. For. Rev.* **2006,** *8* (4), 439–448.

Clawson, M. *Methods of Measuring the Demand for and Value of Outdoor Recreation;* Reprint, No.10, Resources for the Future Inc.: Washington D.C., 1959.

Das, S. *Travel Cost Method for Environmental Valuation;* Centre of Excellence in Environmental Economics, Madras School of Economics: Chennai, India, 2013.

English, D. B. K.; Bowker, J. M. Sensitivity of Whitewater Rafting Consumers' Surplus to Pecuniary Travel Cost Specifications. *J. Environ. Manag.* **1996,** *47* (1), 79–91.

Gössling, S. Ecotourism: A Means to Safeguard Biodiversity and Ecosystem Functions? *Ecol. Econ.* **1999,** *29* (2), 303–320.

Government of India 2002; The Planning Commission of India, Chapter 23. http:// planning commission.nic.in/plans/mta/mta-9702/mta-ch23.pdf. Published online: 2002.

Guha, I.; Ghosh, S. *A Glimpse of the Tiger: How much are Indians Willing to Pay for it?;* South Asian Network for Development and Environmental Economics: Kathmandu, Nepal, 2009.

Gürlük, S.; Rehber, E. A Travel Cost Study to Estimate Recreational Value for a Bird Refuge at Lake Manyas, Turkey. *J. Environ. Manag.* **2008,** *88* (4), 1350–1360.

Kuosmanen, T.; Nillesen, E.; Wesseler, J. Does Ignoring Multidestination Trips in the Travel Cost Method Cause a Systematic Bias? *AJARE.* **2004,** *48* (4), 629–651.

Lindberg, S. I. It's our Time to Chop: Do Elections in Africa Feed Neo-Patrimonialism Rather than Counteract it? *Democratization.* **2003,** *10* (2), 121–140.

McConnell, K. E.; Strand, I. Measuring the Cost of Time in Recreation Demand Analysis: An Application to Sportfishing. *Am. J. Agric. Econ.* **1981,** *63* (1), 153–156.

Mishra, R. *Welfare State Capitalist Society;* Routledge: London, UK, 2014.

Moeltner, K. Addressing Aggregation Bias in Zonal Recreation Models. *J. Environ. Econ. Manag.* 2003, *45* (1), 128–144.

Pedersen, A. *Managing Tourism at World Heritage Sites: A Practical Manual for World Heritage Site Managers;* UNESCO World Heritage Centre: Paris, France 3, 2002; p 96.

Petrosillo, I.; Zurlini, G.; Corliano, M. E.; Zaccarelli, N.; Dadamo, M. Tourist Perception of Recreational Environment and Management in a Marine Protected Area. *Landscape Urban Plan.* **2007,** *79* (1), 29–37.

Prideaux, B.; Falco-Mammone, F. *Economic Values of Tourism in the Wet Tropics World Heritage Area;* Cooperative Research Centre for Tropical Rainforest Ecology and Management: Cairns, Australia, 2007.

Scheyvens, R. Ecotourism and the Empowerment of Local Communities. *Tourism Manag.* **1999,** *20* (2), 245–249.

Silori, C. S. Socio-Economic and Ecological Consequences of the Ban on Adventure Tourism in Nanda Devi Biosphere Reserve, Western Himalaya. *Biodivers. Conserv.* **2004,** *13* (12), 2237–2252.

UNEP and CI (United Nations Environment Programme & Conservation International) 2007. *Tourism and Mountains: A Practical Guide to Managing the Environmental and Social Impacts of Mountain Tours;* United Nations Environment Programme: Nairobi, Kenya, 2007; p 29.

Union, T. W. *United Nations List of Protected Areas;* IUCN: Cambridge, UK, 1998.

Ward, F. A.; Loomis, J. B. The Travel Cost Demand Model as an Environmental Policy Assessment Tool: A Review of Literature. *Western J. Agr. Econ.* **1986,** *11* (2), 164–178.

Willis, K. G.; Benson, J. F. Recreational Values of Forests. *Forestry.* **1989,** *62* (2), 93–110.

Wood, R. E. Ethnic Tourism, the State, and Cultural Change in Southeast Asia. *Ann. Tourism Res.* **1984,** *11* (3), 353–374.

Wunder, S. Ecotourism and Economic Incentives—an Empirical Approach. *Ecol. econ.* **2000,** *32* (3), 465–479.

Zandersen, M.; Tol, R. S. A Meta-Analysis of Forest Recreation Values in Europe. *J. For. Econ.* **2009,** *15* (1), 109–130.

CHAPTER 4

DESTINATION MARKETING APPROACHES FOR WILDLIFE TOURISM

ATHULA CHAMMIKA GNANAPALA*

Department of Tourism Management, Faculty of Management Studies, Sabaragamuwa University of Sri Lanka, Belihuloya, Sri Lanka

E-mail: athulatmsusl@gmail.com

CONTENTS

ABSTRACT

Observing wildlife is one of the major travel motives of tourists and plays an important role in their destination selection process. Tourists like to view non-captured animals in a typical natural setting rather than view the captured animals. Further, a few tourists and travel agents are vocally against the viewing of captured animals. As a result the demand for pure wildlife tourism is growing rapidly, as is competition among wildlife destinations, and destinations are required to have proper and sustainable marketing practices to secure competitive advantage in this situation. This chapter is based on the marketing practices of wildlife tourism destinations, including travel motives, market segmentations, targeting and positioning, marketing, and promotional strategies. The study is mainly based on secondary information, but is supported by empirical evidence from Sri Lanka.

"Wildlife tourism also comes under the leisure category...."

4.1 INTRODUCTION

Wildlife tourism has become one of the fastest growing market segments in the global tourism industry. Higginbottom (2004) defines wildlife tourism as tourism based on encounters with non-domesticated (non-human) animals. These encounters can occur either in the animals' natural environment or in captivity. Fuelling this is the fact that global tourism is increasing very rapidly. A total of 1133 million tourists travelled globally in 2014 and of these 53% (598 millions) travelled for leisure, recreation, and holiday (leisure) purposes (UNWTO, 2015). Wildlife tourism also comes under the leisure category. For that reason, the wildlife segment plays a major role in attracting tourists to destinations in which the government can realize more economic and other advantages. Therefore, the most efficient destinations use effective marketing strategies to attract and serve wildlife tourists efficiently and profitably. According to Kotler and Armstrong (2012, p. 4) marketing, more than any other business function, defines "profitable customer relationships." Therefore, the destination marketers of wildlife tourism need to use two major strategic weapons, firstly attracting new tourists by promising superior value and secondly keeping and growing current wildlife lovers through better quality experiences to meet their expectations.

This chapter is mainly based on the destination marketing approach to wildlife tourism: including what is meant by wildlife tourism; what are the products that can be sold to wildlife tourists?; the market segments of wildlife tourists (to whom to sell?); the relationship between target markets and travel motives (why do people visit wildlife destinations?); destination positioning; and promotional and other strategies for wildlife tourism (how to market/sell wildlife destinations?). The discussion is mainly based on a review of the international literature relating to wildlife tourism. However, even though it focuses on the global perspectives of wildlife tourism, 10 exploratory interviews were conducted with industry stakeholders in Sri Lanka (including academics, travel agents, and wildlife park managers) to provide comparative data. As a result, it is hoped that this chapter will contribute to a sound understanding of the theoretical aspects and practical applications of the marketing principles and strategies best suited for wildlife destinations, so that they will be able to manage themselves efficiently and effectively meet their stated goals.

4.2 THE CONCEPT OF DESTINATION

The term destination is somewhat broad and nebulous (Whittlesea et al., 2015), as it is a geographical space in which a cluster of tourism resources exist (Pike, 2004). The UNWTO (2007) defines a tourism destination as a physical space in which a tourist spends at least one overnight stay, and includes tourism products and services. Further, it includes physical and administrative boundaries, management, images, and market competitiveness. Vengesayi (2003) argues that a destination is a mix of attractiveness and competitiveness, and attractiveness is the ability of the destination to deliver benefits, while competitiveness is the ability of a destination to deliver a better experience than other destinations. Therefore, a destination is just not like a physical place, but is a mix of products and services that are capable of providing more satisfying experiences to tourists (Shaw & Williams, 2004).

Hudson (2008) defines a destination as having physical, political, or even market-created boundaries. Therefore, the tourism destination can comprise a wide range of elements that combine to attract visitors to stay for a holiday or day visit, but there are four core elements that make up the destination product: prime attractors, the built environment, supporting supply services, and socio-cultural dimensions, such as atmosphere or ambience (Lumsdon, 1997). Kotler et al. (2010) argue destinations are a place with some actual or perceived boundary, and also have discussed the concept of macro and micro destinations. A country, for example South Africa, Sri Lanka, or Kenya is a macro wildlife destination, and constituent states, regions, cities, and even a specific wildlife places co-exist with the macro destination as micro destinations.

Attractions are thus one of the most important components in tourism products (Gnanapala, 2015; Hudson, 2008), and they are main travel motivators for visits to a particular (wildlife) destination. In prime attractions, wildlife may a major role in attracting tourists to both micro and macro destinations.

4.3 TRAVEL MOTIVES FOR WILDLIFE TOURISM

Travel motives are the starting point of the buying decision process for any good and services, including wildlife tourism (Crompton & McKay, 1997; Gnanapala, 2012, 2015). According to Uysal and Hagan (1993), motivation is a dynamic concept; it may vary from one person to another, from one segment to another, from one destination to another, as well as from one decision-making process to the next. Murray (1964) defines motive is an internal factor that arouses, directs, and integrates a person's behavior.

As stated by Dann (1981), tourist motivation is a meaningful state of mind which adequately disposes actors to travel. According to Crompton and McKay (1997), travel motivation is a dynamic set of internal psychological factors that generate a state of tension or disequilibrium within individuals. Harmer (2001) defines motives as a kind of internal drives that push someone to do things in order to achieve some benefit. Therefore, internal travel motives pressure tourists to stay in different wildlife destinations and do different activities like safaris, camping, photographing, whale watching, or butterfly watching. The motives may drive wildlife tourists to behave actively or passively, thus, Dornyei (2001) highlighted that motivation is the reason for why people decide to do something, how long they are willing to sustain the activity, and how hard they are going to pursue it.

The concept of push motives are heavily discussed in tourism as identifying travel motives or the reasons for having the holiday (Uysal & Hegan, 1993; Yoon & Uysal, 2003; Snepenger et al., 2006; Fodness, 1994; March & Woodside, 2005; Mayo & Jarvis, 1981; Gnanapala, 2015). Both push and pull motives influence the travel decisions of individuals. These travel motives greatly affect the travel decision as well as destination selection. Oliver (1997) shows that individuals purchase products, for example, a wildlife holiday based on two major motives—to remove an experience deficit and to add something of value for life. Therefore, we can assume that the tourists with similar motives select similar destinations and engage in similar activities. Also, during the destination selection and holiday decision-making process the potential wildlife tourist may have to find viable answers for a number of significant enquiries, including "why do we/I travel?," "where to go?," "with whom to travel?," "which is the most suitable wildlife destination?," "what to do there?," and "when to go?," and the answers to those questions are highly influenced by the nature and the level of the travel motives of the individuals (Gnanapala, 2012, 2015).

Different theories and concepts have been developed to identify and discuss the motivation of individuals generally and particularly in relation to the travel, such as Maslow's hierarchy of needs theory, Iso-Ahola's seeking and escaping theory, Dunn's push & pull motives, McClelland's needs theory, and Clayton Alderfer's ERG theory. Among them push & pull motives and Maslow's hierarchy of need theory have been discussed heavily in the tourism literature due to their high applicability for tourism (Cohen, 1972; Dann, 1977; Crompton, 1979; Gnoth, 1997; Pearce, 1982; Baloglu & Uysal, 1996).

Individuals travel to different wildlife destinations to meet their expectations, and the meeting of these expectations may lead them toward satisfaction. Iso-Ahola (1982) highlighted that people perceive a leisure activity as a potential pleasure producer for two major reasons: escaping and seeking. Those activities may provide certain intrinsic rewards, such as a feeling of mastery and competence, which can help a traveller to escape from the routine environment, and also gratify travel motives, such as to see Leopards in the Sri Lankan national parks or see whales from Sri Lankan beaches (Gnanapala, 2015).

When calculating tourism statistics, especially the purpose of visit, most countries do not include wildlife as a separate category. Wildlife goes under nature or pleasure categories depending on the perception of the tourists, and, therefore, it is very difficult to get exact figures of wildlife tourists. Therefore, there is a need to depend on the statistics of national parks which tourists visit frequently. However, the tourists who engage in wildlife tourism in private locations and other places are not included in official statistics. A lot of wildlife related activities are happening, like birds and butterfly watching and so on outside of public lands, and these are common and under-enumerated activities in most countries when calculating the actual size of the wildlife market segment. Therefore, it is necessary to have separate studies/surveys to identify the exact size/ number of wildlife tourists.

As a result countries and researchers are required to conduct studies to identify the actual size of this segment. According to an airport survey conducted by Sri Lanka Tourism Development Authority (2013), nearly 3% of tourists suggested that their main motive was to visit wildlife. And, when identifying their second and third travel motives, the tourists ranked wildlife as being 6.2 and 11.5%, respectively. Even though the numbers are small as a percentage, industry stakeholders consider wildlife tourism as a lucrative market. According to the country of origin of tourists, Europeans are the major wildlife lovers visiting Sri Lanka, and mainly come from United Kingdom and Germany.

4.3.1 PUSH AND PULL MOTIVES

The concept of push and pull motives was first introduced by Dann (1977) and is considered as a simple and intuitive approach to the travel motivations of tourists. Push and pull factors motivate people to take two different decisions at two different times, that is, whether to go, and where to go. The push motives describe the internal socio-psychological forces which influence an individual to take a holiday (Crompton, 1979; March & Woodside, 2005; Mayo & Jarvis, 1981). The most common push motives that motivate potential tourists are escape from monotonous life or environment, rest and relaxation, social interaction, and/or esteem. Pull factors are the extrinsic motives representing the diverse product and services (diversity of attraction) attributes that attract a person toward a holiday destination (Kassean & Gassita, 2013; Weaver & Oppermann, 2000).

Thus, different studies have identified wildlife as a major factor in the destination selection process of tourists. Yuan & McDonald (1990) identified the push factors as escape, novelty, prestige, enhancement of kinship relationships, relaxation and hobbies, and the pull factors as budget, culture and history, wilderness, ease of travel, cosmopolitan environment, facilities and hunting, with the majority of these motivational factors are associated with the nature and wildlife. Crompton (1979) discussed such push motives as escape, self-exploration and evaluation, relaxation, prestige, regression, and social interactions push motives as novelty and education. Gnanapala (2015) identified that wildlife viewing is a major travel motive in the selection of Sri Lanka as a travel destination by the foreign tourists. The literature also shows that travel motives differ from nation to nation, and Europeans are considered as wildlife enthusiasts, but on the other hand Middle East tourists are comparatively less interested in wildlife.

4.4 MARKET SEGMENTATION IN WILDLIFE TOURISM

As a technical definition, wildlife tourism can be broadly seen to be trips to destinations, with the main purpose of visit being to observe the local fauna (Kurleto, 2014). The major segment of wildlife tourists has safaris in wildlife parks to see wild animals and take some photos, and so on. However, it can be identified that there are many niche markets in the basic wildlife segment, such as butterfly watching, bird leopard safaris, monkey trails, and whale watching. The market of wildlife tourism is growing very rapidly due to the loss of greenery in most city areas, the lack of social contacts in the work place, and the monotonous nature of life. Wildlife tourists may active or passive; however, the majority is passive and they visit wildlife parks as a part of a tour (mass tourism). Their prime motive is not to closely watch wildlife, but be closed to it. Active tourists are the real wildlife lovers, and the main purpose of their visit is the actual wildlife. Their number is small, but it is profitable (this is confirmed by the small industry stakeholder survey, personal communication, 2015). The majority of the pure wildlife tourists are considered as risk takers (Dowd, 2004).

Kotler and Armstrong (2012) define market segmentation as dividing a market into distinct groups with distinct needs, characteristics, or behavior, who might require separate products, or marketing mixes. Similarly, Middleton et al. (2009) characterize segmentation as the process of dividing a total market such as all visitors, or market sectors, such as holiday travel or business travel, into subgroups, or segments for marketing management purposes. Also highlighted is that the main purpose of market segmentation is to build more cost-effective marketing practices through the formulation, promotion, and delivery of purpose-designed offerings that satisfy the exact needs and wants of target customer groups.

The wildlife field is diverse, and it includes many different products and services. In marketing the product or offerings are given a deep meaning. According to Kotler and Armstrong (2012, p. 224), "a product is anything that can be offered to a market for attention, acquisition, use, or consumption that might satisfy a want or need." Therefore, the wildlife product may include tangible products and intangible services, including objects, events, persons, places, organizations, ideas, and so on.

4.4.1 BASES FOR THE SEGMENTING OF WILDLIFE TOURISM

The market of wildlife tourism consists of a range of diverse niche markets. These range from the tourist who seeks excitement in wildlife experiences in unusual environments, to having wildlife tours in a jeep to see animals and take photographs for enjoyment in popular wildlife destinations. Those tourists may considerably differ in their culture, age, gender, income, education, occupation, family life cycle, lifestyle, or personality. Frequently, the wildlife products are sold through middlemen like tour operators and travel agents, and packaged. Considering the needs and wants of the tourists, these intermediaries prepare different tour packages to promote wildlife destinations. Also, they will prepare tailor made tour packages based on special needs and other special requests made by tourists. The larger segment of wildlife tourism thus represents those mass tourists who want to see just animals as an experience rather than for learning or education purposes. Therefore, the mass market may be more passive and tend to create more negative impact on a destination.

The real wildlife lovers are considered to be rich tourists, and to be willing to pay even more to have real wildlife experiences also stay more days in a destination. Wildlife lodgings can in fact be more expensive than five star resorts; furthermore even staying in a campsite may be expensive in Sri Lankan national parks. For example, if the tourist wants to stay in camping sites at Sri Lankan national parks they need to get the support of the service providing organizations with camping equipment; therefore, the cost may be very high. Usually, camping accommodation is considered as an economical mode of accommodation for travellers, however, that at wildlife parks brings more cost since it needs to get the support of many different parties. The pure wildlife tourist may thus be willing to bear the high cost of compensating their expectations of thrilling and exciting experiences.

According to Middleton et al. (2009), market segmentation is the process of dividing a total market, such as all visitors, into subgroups or segments for marketing management purposes. According to them, the main objective of segmentation is to facilitate more cost-effective marketing through the formulation, promotion, and delivery of purpose-designed products that satisfy the identified needs of target groups. If the destination can identify the exact needs of the wildlife tourists, for example, to see leopards or butterflies, or stay in a campsite at a wildlife park, therefore, through market segmentation it may be able to offer a better quality product or services to satisfy the exact needs of the wildlife tourists. However, in mass tourism the destination cannot fulfill such requirements, and this may lead the tourist toward frustration and dissatisfaction.

4.4.2 *BASES FOR SEGMENTING THE WILDLIFE MARKET*

There are many variables that can be used to segment wildlife markets, but all variables may not be equally effective in producing separate market segments. According to Kotler and Armstrong (2012), an effective market segment should consist of the required characteristics, such as measurability, accessibility, substance, differentiation, and be able to be auctioned. Similarly, Middleton et al. (2009) identified four major criteria (discrete, measurable, viable, appropriate, and sustainable) for usable or actionable segments, based on the thoughts of Kotler and Armstrong (2012), and Middleton and Hawkins (1998).

Therefore, there is no hard and fast rule to segment wildlife markets, thus any segment that meets the required criteria can be used to do this. Hence, wildlife destinations can practice different segmentation variables alone or in combinations, to attract the customer groups with different needs to receive the benefit in return. Kotler et al. (2010) identified four major variables to segment consumer markets, and these variables can be equally adopted for any tourism markets, including wildlife. These segmentation variables are geographic (nations, regions, states, counties, cities, or even neighborhoods), demographic (age, gender, family size, family life cycle, income, occupation, education, religion, race, generation, and nationality), psychographic (social class, lifestyle, or personality characteristics), and behavioral characteristics (segments based on their knowledge, attitudes, uses, or responses to a product, such as occasions, benefits, user status, user rates, loyalty status, readiness stage, or attitude toward product). This classification is too general and not specific to tourism or wildlife. However, Middleton et al. (2009) introduced a more specific and practical seven variables exclusively for tourism, such as purpose of travel, buyer needs, motivations, and benefits sought buyer behavior/characteristics of product usage, demographic status, economic and geographic profiles, psychographic profile, geo-demographic profile, and price. Therefore, individual wildlife destinations, through the support of travel intermediaries, such as travel agents and tour operators can prepare exclusive tour packages for the different market segments to provide different wildlife experiences based on a consideration of these variables.

4.4.3 TARGETING MARKETS

During the market segmentation stage the wildlife destinations can identify the possible variables that can be used to segment markets. However, all those markets may not be effective due to various reasons, therefore, destinations need to select the best segment or few segments to reach and serve. According to Kotler et al. (2010), a target market consists of a set of buyers who share common needs or characteristics that the company decides to serve. Further, they have highlighted that the market targeting can be carried out at several different levels, such as differentiated marketing, undifferentiated marketing, concentrated marketing, and micromarketing. When selecting a market segment, the marketers need to evaluate each segment using three important criteria, such as segment size and growth, structural attractiveness of the segment, and company objective and resources. The destinations can go for the best segment/s depending on their unique characteristics and the availability of resources. In wildlife tourism organizations can practice even micromarketing (local or individual marketing) successfully. Wildlife destinations can serve both locals and international tourists, and even charge two different prices. Similarly, the tourists' interests and the requirements may also differ from segment to segment.

4.4.4 POSITIONING

After identification of the target markets and customer groups organizations can develop better quality tour packages to attract the targeted tourists. However, the tourists will receive a lot of information regarding different wildlife destinations, and it may be that potential tourists as consumers are overloaded with information about the different wildlife destinations and related products, services, and other offerings. Therefore, the consumers may face difficulties to evaluate the products and other offerings when they make buying decisions. Consequently, destinations should position their offerings in the consumer markets efficiently and effectively. The purpose of positioning is to create certain images in the minds of potential customers (Dimanche & Sodja, 2007; Eraqi, 2007). However, positioning is more than just image creation, and it aims to establish the image and create a competitive edge to the brand, product, or destination (Hooley et. al., 2004).

According to Kotler and Armstrong (2012, p. 207), "a product position is the way a product is defined by consumers on important attributes, the place the product occupies in consumers' minds relative to competing products." Similarly, Lovelock (1991) defines positioning as establishing and maintaining a distinctive place in the market for an organization and/or its individual product offerings. However, according to McDonald et al. (2001), positioning is not about what you do to the product, but what you do to the customer, and how the customer perceives you. As highlighted by Hooley et al. (2004), positioning is thus more than just image creation among the potential target groups, and it aims to establish the image and create a competitive edge to the brand, product or destination.

Selecting an appropriate positioning strategy is an essential but difficult task for wildlife marketers. The positioning strategy with a proper message should be able to stimulate potential buyers. Kapferer (1997) highlights that it needs to answer four questions when creating an effective positioning strategy which can add value to the positioning process. These are: why? (this refers to the brand promise and consumer benefit aspect); for whom? (refers to the target aspect); when? (on what occasions will the product be consumed?); and against whom? (who are the main competitors, what clients/customer the organization thinks they can conquer?). Aaker et al. (1996) argue that the process of positioning includes both products and services, and how they are communicated to the target market and emphasized in order to keep a narrow focus. Positioning is then not aiming at wider markets in order to avoid creating a fuzzy image of the destination. This statement is highly applicable to those destinations that want to promote wildlife tourism to attract more tourists.

According to Kotler and Armstrong (2012), differentiation and positioning are interrelated, and the task consists of three steps. These are: first, identifying a set of differentiating competitive advantages on which to build a position; second, choosing the right competitive advantage approach; and finally, selecting an overall positioning strategy. Positioning is also focused on competition as the customers compare and keep in their minds how the product/brand is similar or different from competing products. Therefore, a destination should create a position that takes into consideration not only the destination's own strengths and weaknesses, but also the strengths and weaknesses of competitors (Blankson & Kalafatis, 1999b).

Wildlife destinations must then effectively communicate and deliver this chosen position to the market to build profitable relationships with target customers, and marketers must understand tourists' needs better than competing destinations do, to deliver more value to them. As highlighted by Kotler and Armstrong (2012, p. 210), "to the extent that a company can differentiate and position itself as providing superior customer value, it gains competitive advantage." Porter (1996) defines the two types of competitive advantage an organization can achieve relative to its rivals; lower cost and/or differentiation. Kotler et al. (2010) shows that an advantage over competitors is gained either by having lower prices, or by providing more benefits that justify higher prices. It is known that wildlife tourism products and services are more expensive than other products and services, and, therefore, it is necessary to go for differentiation of wildlife products using unique attractions and associated benefits. However, tourism destinations, especially in developing countries, can get these advantages largely though the lower prices of their supplementary and complimentary offerings, such as food and beverages, accommodation, and other related products and services. Therefore, adopting proper planning and strategies will enable a destination to position wildlife tourism products effectively by targeting potential tourists.

4.4.5 POSITIONING STRATEGIES IN WILDLIFE TOURISM

The responsibility of promoting wildlife tourism often first resides with the National Tourism Organization (NTO) of a particular country, with the support of the relevant stakeholders in wildlife tourism, including wildlife parks, forests management, government, local governments, tour operators, and travel agencies. Therefore, it is necessary to form destination-marketing organizations (DMOs) to position the destination, together with wildlife resources including whales, leopards, elephants, birdlife, butterfly, and other fauna varieties. According to McCabe (2009), in each destination the DMO must communicate the diversity of attractions to appeal to a wider set of market segments, and must communicate something of the core attributes or benefits of the whole region while doing this.

Wildlife destinations are required to use different and effective positioning strategies to attract tourists due to the ever-increasing competition between countries, as well as within the countries between different wildlife parks. Hence, it is necessary have strong positioning strategies, called value positioning, for wildlife tourism destinations. According to Kotler and Armstrong (2012, pp. 212–213), their value proposition (the full positioning of a brand) is the full mix of benefits on which a wildlife destination is differentiated and positioned, and is the answer to the customer's question "why should I buy your brand?"

Therefore, the different positioning strategies or themes can be used by the DMOs in wildlife destinations, such as attribute positioning (highlights the most common and attractive animals like whales, leopard, elephants, and bears in Sri Lanka), benefit positioning (highlights the exclusive benefits the tourists can consume), use/application positioning (highlights that the park is ideal for certain types of activities), user positioning (the park is ideal for a certain types of wildlife tourists, e.g., young adults or empty nesters), competitor positioning (the wildlife reserve is richer than competing destinations), product category positioning (a park is ideal for leopards), and quality/price positioning (providing the best value for money).

Finally, destinations need to develop an effective and attractive positioning statement, which can be used in promotional materials. A positioning statement is a concise description of the target market as well as a persuasive picture about the destination, and how the destination wants to position its offerings for potential tourists. As highlighted by Kotler and Armstrong (2012, p. 215), the positioning statement should follow the form: to (target segment and need) our (brand) is (concept) that (point of difference).

4.4.6 DESTINATION BRANDING

Destination positioning is closely interrelated with destination branding in relation to wildlife tourism. According to Kotler and Armstrong (2012, p. 231) "a brand is a name, term, sign, symbol, or design, or a combination of these, that identifies the maker or seller of a product or service." Doyle (1989) discussed the term branding from the producers perspective rather than the consumers perspective, and defines a brand as a name, symbol, design, or some combination of these, which identifies the product of a particular organization as having a sustainable differential advantage. Similarly, Hudson (2008) defines branding as a method of establishing a distinctive identity for a product, based on competitive differentiation from other products. The tourist may view a brand as an important part of a wildlife offering, and branding can add more value to a destination. Customers attach meanings to brands and develop brand relationships and may go beyond a product's tangible offerings. As highlighted by Hudson (2008), destination branding has received increased attention over the last few decades due to the ever-increasing global competition for tourists. Therefore, destinations including wildlife ones need to build and maintain a distinctive identity to distinguish themselves from other competitors, and this has become more critical in the present digital era. Branding a wildlife destination that creates a superior proposition, that is, distinctive from competitors will bring more competitive advantage and other intangible benefits.

In addition to the implicit advantages that derive from brand positioning, Middleton and Clarke (2001, pp. 133–134) suggest that of tourism products and destinations wildlife branding brings more specific advantages. These are: first, that it helps reduce medium and long-term vulnerability to the unforeseen external events that so beset the tourism industry; second, it reduces risk for the consumer at the point of purchase by signaling the expected quality and performance of an intangible product; third, it facilitates accurate marketing segmentation by attracting some consumer segments and repelling others; fourth, it provides the focus for the integration of stakeholder effort, especially for the employees of an organization, or the individual tourism providers of a destination brand; fifth, it serves as a strategic weapon for long range planning in tourism, and finally, many see clearly recognized international branding as an essential attribute for effective use by businesses of the communication and distribution abilities of the Internet.

The countries rich in biodiversity and wildlife as well as in the major pull factors based on wildlife can easily go for destination marketing and branding as a pure wildlife destination (e.g., Kenya, Tanzania, Botswana, South Africa, and Namibia). On the other hand, the countries which have a diversity of attractions *including* wildlife, can position themselves as a compact destination (e.g., Sri Lanka, Malaysia, and Thailand). There may also exist several popular micro destinations within a popular macro destination (Kotler et al., 2010). Therefore, branding and marketing should be done collaboratively to enjoy their mutual benefits and to reduce malpractices, such as over positioning. According to the Lonely Planet (2015), the top 10 ultimate places for wildlife are Belize, Bolivia, Botswana, the Great Barrier Reef-Australia, Costa Rica, the Everglades—USA, Kenya, Galapagos Island-Ecuador, Madagascar, and Malaysian Borneo. This list contains both main destinations/countries as well as micro destinations (specific tourist attractions within countries). Travelers Digest (2015) also has ranked the top 10 wildlife destinations of the world as: Botswana, Galapagos Islands, Amazon Rainforest, Alaska, Rwanda, Antarctica, Madagascar, Shaanxi Province-China, Churchill Manitoba of Canada, and Yellowstone National Park, USA. Similarly, the top 10 popular wildlife destinations in Asia are Gunung Leuser National Park (Indonesia), Bonin Islands (Japan), Danum Valley Conservation Area (Sabah, Malaysia), Calauit Wildlife Sanctuary (Philippines), Similan Islands (Thailand), Yala National Park (Sri Lanka), Xe Pian National Protected Area (Laos), Shaanxi Province (China, Woraksan National Park (South Korea), and Ranthambore National Park of India (Travelers Digest, 2015).

Since most renowned travel guides recommend these countries and places as the most popular wildlife destinations, they benefit from favorable publicity and the development of strong brand beliefs, and motivation among potential tourists. Most of the travel related magazines, guides and other media declare that tourists *should* visit the wildlife destinations of the world. However, there are no clearly visible criteria and guidelines to select the best and most popular destinations, and the selection method is, therefore, more subjective. Nevertheless, a destination can obtain favorable publicity since these media are both popular credible information channels used among the target wildlife customer groups.

Most of the popular wildlife tourist destinations lie in the same country or the same region; therefore, it is often necessary to go for collaborative marketing and branding (CRC for Sustainable Tourism, 2012). This will encourage sustainable marketing practices, and it also encourages a destination to go into conservation and protecting wildlife with greater responsibility. Also, such co-marketing will encourage tourists to visit a region rather than to stay in one destination, and all the stakeholders will benefit, including tourists. The tourists will be able to enjoy different wildlife resources, enabling them to receive more value for money, and finally, more visitor satisfaction.

4.5 PROMOTIONAL STRATEGIES

In the present competitive tourism industry, information, in terms of effective communication, plays a major role for the success or failure of business organizations including those that deliver wildlife tourism. As highlighted by Kotler and Armstrong (2012, p. 407) "building good customer relationships calls for more than just developing a good product, pricing it attractively, and making it available to target customers." The wildlife destinations/marketers need to communicate their value propositions to customers, and all communications must be planned and blended into carefully integrated programs. Therefore, it is necessary to develop an effective promotional mix for wildlife destinations, to attract tourists and satisfy them in order to achieve the objectives of the destination.

According to Kotler and Armstrong (2012), a company's total promotion mix consists of the specific blend of advertising, public relations, personal selling, sales promotion, and direct-marketing tools that a company uses to persuasively communicate customer value and build customer relationships. The following section briefly discusses the major promotion tools:

- ***Advertising*** is any paid form of non-personal presentation and promotion of ideas, goods, or services by an identified sponsor (includes broadcast, telecast, print, internet, outdoor, and any other forms);

- ***Sales promotion*** is the providing of short-term incentives for tourists, middlemen and sales force to encourage the purchase or sale of a product or service (includes discounts, coupons, displays, and demonstrations);
- ***Personal selling*** is the personal presentation by the firm's sales force for the purpose of making sales and building customer relationships (includes sales presentations, trade shows, and incentive programs);
- ***Public relations*** is the building of good relations with the company's various publics by obtaining favorable publicity, building up a good corporate image, and handling or heading off unfavorable rumors, stories, and events (includes press releases, sponsorships, special events, and web pages); and
- ***Direct marketing*** is the direct connections with carefully targeted individual consumers to both obtain an immediate response and cultivate lasting customer relationships (includes telephone marketing, the Internet, and mobile marketing).

Even though there are many promotional tools and techniques available for wildlife marketers, all those techniques are always effective due to the nature of the product and the special characteristics of wildlife services, such as intangibility, inseparability, perishability, and variability. Therefore, marketers need to select the most appropriate and cost effective promotional tools. The tools that can touch more of the senses of the tourists will be more effective, and through these tourists will be able to receive more persuasive appeals to create positive attitudes and perceptions about the wildlife destination. The tourist needs to use all his/her senses to receive the information provided by wildlife marketers. However, there are some advantages and disadvantages in all the media used to achieve this. Print-based media, for example, newspapers and magazines can provide information, and clear and exciting pictures to get the attention and awareness of wildlife enthusiasts. Film-based advertisements can provide more information of natural environments, and will touch more senses to arouse the travel motives of the potential tourists. However, direct marketing and social media marketing bring more positive impacts to tourism destinations than other promotional tools. In the global context, social media marketing is becoming a more important and effective promotional media vehicle in the tourism and hospitality industry.

According to Middleton et al. (2009), public relations are a powerful media tool in tourism. Therefore, wildlife destinations can use media relations wisely and effectively to get favorable and mutual comparative advantage. Media relations are about obtaining non-paid-for media coverage, and the activity includes writing press releases, feature articles, scripts, preparing press packs, obtaining interviews, holding press conferences, and creating an up to date database of media contacts. Since customers receive information from a third party the credibility and believability may be very high, and is obtained at zero cost. As highlighted by Middleton et al. (2009), through regular contact with journalists who specialize in travel or cover the local catchment area of the organization, a wildlife destination that can develop good relations can build up trust and make it easier to get the organization's messages across when it matters.

4.5.1 TRADE FAIRS AND EXHIBITIONS

Trade fairs and exhibitions are the kind of events that can deliver effective marketing in the tourism and hospitality industry. Events bring tourists to a place where they can see, touch, listen, smell, and sometimes taste the goods on display. The trade fairs and exhibitions are largely used and companies invest in them since only direct selling receives more funds than fairs in terms of the marketing mix. Participation at trade fairs and exhibitions will bring opportunities and get the attention of both the tourists and the middlemen in the tourism and hospitality industry. As highlighted by Neven and Kanitz (2013), communication is the act of exchanging messages and information, therefore, when a company takes part in a trade fair information is exchanged. Trade fairs and exhibitions thus provide an opportunity to have real-life encounters between buyers and sellers and achieve greater clarification of a product's appeal.

Many tourism and hospitality organizations exhibit their offerings at travel trade shows, exhibitions or conventions, and different stakeholders of the industry, such as suppliers, carriers, intermediaries, and destination-marketing organizations participate together. Middleton et al. (2009) argue that the arrangement of trade fairs and exhibitions is similar with regard to all the variables of the promotional mix. For example, prior to an event exhibitor often send public and personal invitations to regular customers, intermediaries, and general public inviting them to visit their booths. The participants (exhibitors) display their (wildlife) products and other offerings either in print or in digital media form as a type of advertising campaign. Even during the exhibition the sales force of the respective suppliers contact participants through answering questions, and inquiries will also attract leaflets, brochures, business cards, and other tangible evidence as a means of personal selling. The participants will be able to be involved in various sales promotional activities used by the suppliers, and these also give various incentives to encourage buying decisions of the consumers. During the exhibition period, suppliers collect the contact information of participants, and during the post conference period the respective companies contact them through direct mails, telemarketing, and other social media. A wildlife destination can participate in leading global trade fairs like FITUR-Madrid (Spain), REISEN-Hamburg (Germany), BIT-Milan (Italy), ITB-Berlin (Germany), MITT-Moscow (Russia), BITTM-Beijing (Chin), and WTM-London (United Kingdom), with their tangible evidence to create strong appeal to the psychological feelings of the potential wildlife lovers.

Trade fairs and exhibitions provide key major advantages when compared with other promotional tools; that is, the ability to have face to face contacts with buyers, sellers, and other key stakeholders (Reychav, 2009; Palumbo & Herbig; 2002; Shoham, 1992; Kirchgeorg et al. 2010), provides an opportunity for different parties to meet on neutral ground (Reychav, 2009; Shoham, 1992; Skov, 2006), and provide opportunity for people to get first-hand information and experience the branding (wildlife) products and services (Shoham, 1992; Kirchgeorg et al., 2010).

4.5.2 FAMILIARIZATION (FAM) TOURS

Familiarization (FAM) tours are considered as one of the most cost effective promotional tools in tourism and hospitality industry. Wildlife destinations use FAM tours to get the attention of potential wildlife tourists and other key stakeholders. Basically a FAM tour involves a destination attraction hosting tour operators, travel agents, and travel media seeking to create awareness among them, and get a favorable comment from them later (Middleton et al., 2009; McCabe, 2009). According to Middleton et al. (2009), through FAM tours foreign travel agents, journalists, and tour operators visit destinations and sample the products available. NTOs need to play a great role in facilitating these tours, with other key stakeholders of the destination also influencing the effectiveness with which the travel trade in markets of origin acts in support of a destination and its products.

FAM trips are critical to overcome intangibles in tourism. A destination will be able to get tangible and intangible benefits through effectively planned FAM tours. The participants in the tour, such as media personnel, travel agents, tour operators, and other parties, tell their experiences though different mediums to their immediate target customers in numerous ways. For example, tour operators and travel agents will prepare wildlife tour packages and sell to tourists. Similarly, media organizations may prepare video programs and telecast in their channels or write newspaper or magazine articles to inform and persuade the buying motives of wildlife tourists. Most of wildlife destinations especially invite media teams to come and prepare documentaries and telecast them. The most important of these globally are TV channels like National Geographic, Animal Planet, and Discovery, and they play a major role in promoting wildlife tourism globally through educating their clientele. Also, the participants of the FAM trips are important since they also an ability and capacity to act as opinion leaders to promote wildlife tourism services successfully.

4.5.3 SOCIAL MEDIA MARKETING AND BLOGS

According to Buhalis and Foerste (2015), advanced technology enables users to amalgamate information from various sources on their mobile devices, personalize their profile through applications and social networks, and interact dynamically with their context. Therefore, in the present digital business world social media and networking sites allow people to construct their own online communities by linking their personal page to those of their friends and relatives. Social media marketing refers to the process of gaining website traffic or attention through social media sites, such as Facebook, Twitter, LinkedIn, Google Plus, and You Tube. A destination appeal spreads from user to user and presumably resonates because it appears to come from a trusted, third-party source, as opposed to the brand, or company itself. Hence, this form of tourism marketing is driven by word of mouth (WOM), meaning it results in earned media rather than paid media. Social media has become a platform, that is, easily accessible to anyone with Internet access. Additionally, social media serves as a relatively inexpensive platform for organizations to implement marketing campaigns. Most of the tourism related business organizations also has created their social media sites to communicate with their existing and potential clients.

The tourists who visit a particular wildlife destination will express his/her experience with evidence like photographs or video clips also serve as testimonials in different travel blogs. Through social media, the individuals can spread both favorable and unfavorable information about the destination. However, the business firm cannot control the spreading of negative information; the only thing marketers can do is to provide quality services to satisfy or delight the customers. Satisfied customers will tell their positive experiences to third parties, but even more so will unsatisfy customers.

4.6 CONCLUSION

Wild life tourism is watching wild animals in their natural habitat and is a part of eco and nature-based tourism. According to Higginbottom (2004), wildlife tourism is based on encounters with non-domesticated animals and encounters can occur either in the animals' natural environment or in the captivity. Wildlife tourism has become an important segment of the tourism industry in many countries, both developed and developing. Presently, the demand for wildlife tourism has increased dramatically as a result of the growing demand for nature and environment friendly vacations. Therefore, wildlife destinations devote much effort to attract more tourists to their destinations, and as a result the competition also has mounted at both macro and micro destination levels. Destinations need to adopt and practice effective marketing and promotional strategies to face this competition successfully to get competitive advantages. Therefore, wildlife destinations require: first, to identify the exact travel needs (motives) of tourists; second, based on the travel needs it is necessary to identify the effective market segments to develop better quality products and services to fit with the exact needs of the wildlife enthusiast; and finally destinations need to position their offerings in the potential wildlife tourist mind in the form of effective marketing and promotional strategies.

Different promotional methods are widely available for wildlife marketers to position and promote their destinations, such as advertising, sales promotions, personal selling, public relations, and direct marketing tools. All those tools are traditionally practiced by marketers to promote their goods and services. As far as wildlife marketing is concerned, the tools that can target a greater number of the senses of the target customer groups, including travel intermediaries, may be much more cost effective. Therefore, in addition to the traditional promotional tools like advertisements, sales promotions and public relations, non-traditional, and ICT-based strategies are considered as cost effective promotional tools. For example, participation in trade fairs and exhibitions, FAM tours including media tours, social media marketing including travel blogs, and ICT-based direct marketing are important. Even though a destination can utilize various promotional strategies to promote their wildlife resources, their ultimate success is dependent on the ability of destinations to fulfill the expectations of the wildlife tourists that are promised though such promotional activities. The satisfied tourists will behave positively, that is, recommend the destination to their friends and relatives, promote revisits, spread the positive WOM comments, and express their experiences in travel blogs, and so on. WOM is considered as the most powerful and effective marketing strategy as far as destination marketing (wildlife tourism) is concerned, since the information is seen to come from more credible and believable personal sources.

KEYWORDS

- **destination marketing**
- **wildlife tourism**
- **travel motives**
- **positioning**
- **marketing strategies**

REFERENCES

Aaker, D. A.; Myers, J. G.; Batra, R. *Advertising Management;* Prentice Hall: Upper Saddle River, NJ, 1996.

Baloglu, S.; Uysal, M. Market Segments of Push and Pull Motivations: A Canonical Correlation Approach. *Int. J. Contemp. Hosp. Manag.* **1996,** *8* (3), 32–38.
Blankson, C.; Kalafatis, S. P. Issues of Creative Communication Tactics and Positioning Strategies in the UK Plastic Card Service Industry. *J. Market. Commun.* **1999b,** *5,* 55–70.
Buhalis, D.; Foerste. M. SoCoMo Marketing for Travel and Tourism: Empowering Co-Creation of Value. *JDMM.* **2015,** *4* (3), http://dx.doi.org/10.1016/j.jdmm.2015.04.001.
Cohen, E. Towards a Sociology of International Tourism. *Soc. Res.* **1972,** *39,* 164–182.
Crompton, J. I. Motivations for Pleasure Vacations. *Ann. Tour. Res.* **1979,** *6* (4), 408–424.
Crompton, J. L.; McKay, S. L. Motives of Visitors Attending Festival Events. *Ann. Tour. Res.* **1997,** *24* (2), 425–439.
Dann, G. M. Anomie Ego-enhancement and Tourism. *Ann. Tour. Res.* **1977,** *4* (4), 184–194.
Dann, G. M. Tourism Motivations: An Appraisal. *Ann. Tour. Res.* **1981,** *8* (2), 189–219.
Dimanche, F.; Sodja, M. Destination Image and Positioning: The Role of Sports. In *Management Destination Marketing,* March 2007, pp 51–52.
Dörnyei, Z. *Teaching and Researching Motivation*; Longman: Harlow, England, 2001.
Dowd, J. Risk and the Outdoor Adventure Experience: Good Risk, Bad Risk, Real Risk, Apparent Risk, Objective Risk, Subjective Risk. *AJOE.* **2004,** *8* (1), 69–70.
Doyle, P. *Marketing Management and Strategy;* Prentice-Hall: Englewood Cliffs, NJ, 1994.
Eraqi, M. I. Egypt as a Macro Tourist Destination: Tourism Services Quality and Positioning. *IJSOM.* **2007,** *3* (3), 297–315.
Fodness, D. Measuring Tourism Motivation. *Ann. Tour. Res.* **1994,** *21* (3), 555–581.
Franzen, G.; Bouwman, M. *The Mental World of Brands: Mind, Memory and Brand Success;* World Advertising Research Centre: Henley-on-Thames, UK, 2001.
Gnanapala, W. K. A. C. *Travel Motives, Perception and Satisfaction;* Scholar's Press: Saarbrucken, Germany, 2015.
Gnanapala, W. K. A. C. Travel Motivations and Destination Selection: A Critique. *IJRCM.* **2012,** *2* (1), 49–53.
Gnoth, J. Tourism Motivation and Expectation Formation. *Ann. Tour. Res.* **1997,** *24* (2), 283–304.
Harmer, J. *The Practice of English Language Teaching;* Longman Press: Essex, UK, 2001.
Higginbottom, K. Wildlife Tourism: An Introduction. In *Wildlife Tourism: Impacts, Management and Planning;* Higginbottom, K., Ed.; Common Ground Publishing in Association with the CRC for Sustainable Tourism: Gold Coast, Australia, 2004; pp1–14.
Hooley, G.; Saunders, J.; Piercy, N. *Marketing Strategy and Competitive Positioning*; Prentice Hall: London, 2004.
Hudson, S. *Tourism and Hospitality Marketing: A Global Perspective;* Sage: London, 2008.
Iso-Ahola, S. E. Toward Social Psychology Theory of Tourism Motivation: A Rejoinder. *Ann. Tour. Res.* **1982,** *9* (2), 256–262.
Kapferer, J. N. *Strategic Brand Management: Creating and Sustaining Brand Equity Long Term*. Kogan Page Ltd.: London, 1997.
Kassean, H.; Mauritius, R. Exploring Tourists Push and Pull Motivations to Visit Mauritius as a Tourist Destination. *AJHTL.* **2013,** *2* (3), 1–13.
Kirchgeorg, M.; Jung, K.; Klante, O. The Future of Trade Shows: Insights from a Scenario Analysis. *J. Bus. Ind. Mark.* **2010,** *25* (4), 301–312.

Kotler, P.; Armstrong, G. *Principles of Marketing;* Pearson Prentice Hall: Upper Saddle River, NJ, 2012.
Kotler, P.; Bowen, J. T.; Makens, J. C. *Marketing for Hospitality and Tourism;* Pearson Education: Upper Saddle River, NJ, 2010.
Kurleto, M. Managing the Wildlife Tourism Experience: Protecting of Wild Animals and Safeguarding of the Tourists. *J. Bus. Econ.* **2014,** *5* (8), 1403–1412.
Lonely Planet 2015. http://www.lonelyplanet.com/travel-tips-and-articles/57576 retrieved on 16-09-2015.
Lovelock, C. H. *Services Marketing;* Prentice Hall: Upper Saddle River, NJ, 1991.
Lumsdon, L. *Tourism Marketing;* Thomson Business Press: Oxford, UK, 1997.
March, R. G.; Woodside, A. G. *Tourism Behaviour: Travellers' Decisions and Actions;* CABI Publishing: Cambridge, UK, 2005.
Mayo, E.; Jarvis, L. *The Psychology of Leisure Travel: Effective Marketing and Selling of Travel Services;* CBI Publishing Co., Inc.: Boston, MA, 1981.
McCabe, S. *Marketing Communications in Tourism and Hospitality;* Butterworth-Heinemann: Oxford, UK, 2009.
McDonald, M.; de Chernatony, L.; Harris, F. Corporate Marketing and Service Brands: Moving Beyond the Fast-Moving Consumer Goods Model. *Eur. J. Mark.* **2001,** *35,* 335–353.
Middleton, V.; Clarke, J. *Marketing in Travel and Tourism.* MPG Books Ltd.: Rochester, Kent, 2001.
Middleton, V. T. C.; Fyall, A.; Morgan, M.; Ranchhod, A. *Marketing in Travel and Tourism;* Butterworth- Heinemann: Oxford, UK, 2009.
Murray, E. J. *Motivation and Emotion;* Prentice Hall: Englewood Cliffs, NJ, 1964.
Neven, P.; Kanitz, S. *Successful Participation in Trade Fairs;* Auma: Berlin, Germany, 2013.
Oliver. R. L. *Satisfaction: A Behavioral Perspective on the Consume;* Irwin/McGraw-Hill: New York, NY, 1997.
Palumbo, F.; Herbig, P. A. Trade Shows and Fairs: An Important Part of the International Promotion Mix. *J. Promot. Manag.* **2002,** *8* (1), 93.
Pearce, P. Perceived Changes in Holiday Destinations. *Ann. Tour. Res.* **1982,** *9* (2), 145–164.
Pike, S. *Destination Marketing Organisations;* Elsevier: Oxford, UK, 2004.
Porter, M. E. What is Strategy? *Harvard Bus. Rev.* **1996,** *74* (6), 61–78.
Reychav, I. Knowledge Sharing in a Trade Show. A Learning Spiral Model. *J. Inf. Knowl. Manag. Syst.* **2009,** *39* (2), 143–158.
Shaw, G.; Williams, A. M. *Tourism and Tourism Spaces;* SAGE: London, 2004.
Shoham, A. Selecting and Evaluating Trade Shows. *Ind. Market. Manag.* **1992,** *21,* 335–341.
Skov, L. The Role of Trade Fairs in the Global Fashion Business. *Curr. Sociol.* **2006,** *54* (5), 764–783.
Snepenger, D.; King, J.; Marshall, E.; Uysal, M. Modelling Iso-Ahola's Motivation Theory in the Tourism Context. *J. Travel Res.* **2006,** *45* (2), 140–149.
Sri Lanka Tourism Development Authority 2013; Survey of Departing Foreign Tourists from Sri Lanka 2013, SLTDA: Colombo, Sri Lanka, 2013.
Sustainable Regional Tourism Destinations 2010; Best Practice for Management, Development and Marketing. CRC for Sustainable Tourism Pty Ltd.: Queensland, Australia, 2010.

Travelers Digest 2015. http://www.travelersdigest.com/607-wildlife/2/ retrieved on 16-09-2015.
UNWTO2015; Tourism Highlights. UNWTO: Madrid, Spain, 2015.
UNWTO2007; Towards Measuring the Economic Value of Wildlife Watching Tourism in Africa – Briefing Paper, UNWTO: Madrid, Spain, 2007.
Uysal, M.; Hagan, L. R. Motivation of Pleasure to Travel and Tourism. In *Encyclopedia of Hospitality and Tourism;* Khan, M., Olsen. M., Var, T., Eds.; Van Nostrand Reinhold: New York, NY, 1993; pp 798–810.
Vengesayi, S. In *Destination Attractiveness and Destination Competitiveness: A Model of Destination Evaluation,* ANZMAC 2003, Conference Proceedings, Dec1–3, 2003; Monash University: Adelaide, Australia, 2003, 637–645.
Weaver, D.; Oppermann, M. *Tourism Management;* Wiley: Milton, Australia, 2000.
Whittlesea, E.; Hurth, V.; Agarwal, N. Greening the High-spend Visitor: Implications for Destination Marketing. In *Tourism in the Green Economy;* Reddy, M. V., Wilkes, K., Eds.; Routledge: London, 2015.
Yuan, S.; McDonald, C. Motivational Determinants of International Pleasure Time. *J. Travel Res.* **1990,** *29* (1), 42–44.

CHAPTER 5

WILDLIFE TOURISM: TECHNOLOGY ADOPTION FOR MARKETING AND CONSERVATION

AZIZUL HASSAN[1*] and ANUKRATI SHARMA[2]

[1]*Cardiff Metropolitan University, Cardiff, Wales, UK*

[2]*Department of Commerce and Management, University of Kota, Rajasthan, India*

E-mail: M.Hassan15@outlook.cardiffmet.ac.uk

CONTENTS

ABSTRACT

Citing some key examples from selected parts of the world, this conceptual chapter concentrates on technology application for both promotion and conservation of wildlife tourism. Also, the chapter sheds light on relatively two unexplored and less heard wildlife tourism destinations from Bangladesh and India. Examples from Bangladesh and India are expected to visualize the context where, technology application can essentially bring changes of mostly positive types. Wildlife tourism is said to get the attention of academics, researchers, and benefits groups across the world. The obvious reason is its appeal to special tourist groups. This chapter validates the application of appropriate technologies to leave positive impacts on wildlife tourism. It suggests that wildlife tourism needs to capitalize the perceived potentials of online networking readily available on the Internet for wildlife tourism branding and marketing in the global market place. This type of tourism also needs to be aligned with specially developed technological innovations for conservation initiatives in wildlife tourism. In particular, the application of unique technologies is much needed for safeguarding and protection of wildlife when, illicit activities are turning as a backdrop of wildlife tourism. This chapter finds a congenial interrelationship between special technology application and the generic betterment of wildlife tourism. This chapter thus concludes by exploring the basic requirements of useful technology application while stressing on allowing wildlife species to move freely and safely to support tourism activities.

"Wildlife tourism also shows equal growth prediction...."

5.1 INTRODUCTION

Wildlife tourism is not a recent concept, rather it is rooted in hundreds of years of history. Still, wildlife tourism is relatively less explored in tourism literature. Wildlife tourism is seen as both creature- and nature-friendly. Theoretically, wildlife tourism is the watching of wildlife and its habitat in the usual environment. This tourism type in recent years has experienced increasing popularity by adding features such as sustainability and environmental friendliness. Tourism is the single largest industry in the world, contributing at least US$ 3.6 trillion in the global economy amounting to 8% of the global jobs (WTTC, 2014). This form of tourism is relatively popular in many countries across the world having abundance in jungles and wildlife resources. These countries include Australia, South American and African countries, Canada, India, Indonesia, Bangladesh, and others. Following the existing trend of popularization as a process, the contribution of wildlife tourism is also increasing in these countries.

It is evident that tour operators are increasingly relying on technology application for promotional branding and marketing activities for particular products or services (Dadwal & Hassan, 2015; Azim & Hassan, 2013; Hassan, 2012a). Many specialized tour operators across the world are organizing wildlife tours. Wildlife-watching or safari tours are popular in such context that aims both highlighting and promoting wildlife resources before the audiences. In common, both wildlife-watching and safari tourism are designed and organized to watch the real wildlife and have dramatically expanded in markets over the last few years by tour operators through customized tour packages. As the contribution and popularity of wildlife tourism increases, this also threatens the natural habitat of wildlife leading to changing behavior, living patterns, and living spaces. By nature, wildlife species have a particular vulnerability to human disturbances during both their breeding and offsprings juvenile time. Such disturbances may lead to disrupted mating or courtship behaviors, breeding, and offspring care. In wildlife-watching tourism, tourists are very often taken closer to offspring—mother groups thus limiting their caring capacities and disturbing their natural development. Following excessive interruption of visitors, cubs can be separated from their mothers and can become vulnerable to predators. In some cases, wildlife tourism may have serious ill-effects relating to the survival of particular wildlife species. In addition, the illicit activities of poachers threaten wildlife species, limiting the scopes of wildlife tourism activity expansion. Since wildlife tourism is conceptually attached to both marketing and wildlife disturbances, alterations need to be explored aiming to reduce the negative effects to promote wildlife tourism to the best (Green & Giese, 2004). This chapter outlines diverse aspects of wildlife tourism, covering theoretical rhetoric about its definition and the use of technology for marketing and conservation. A series of relevant literature is cited to support the arguments while the chapter critically discusses the application of technology in wildlife tourism for marketing and conservation. As a unique feature, cases from Bangladesh and India are presented as unexplored wildlife tourism destinations to see how technology can support the cases in the given contexts.

5.2 WILDLIFE TOURISM—DEFINITION AND DIVERSITIES

Wildlife tourism is believed to focus on both flora and fauna as important biodiversity components. However, particular focus is mostly placed on fauna only rather than on flora. Generally, wildlife tourism refers to tourism activities with wildlife in the wilderness. Particularly, wildlife tourism can be as simple as viewing a flock of wild birds or a group of large mammals or a bit more complex as tourism activities related to all animal types, insects, and marine life (Tapper, 2006). The World Tourism Organization (UNWTO) (2015) recognizes tourism as a cultural, social, and economic phenomenon outlining peoples' movement to places or countries outside their usual environment for business, professional, or personal reasons. A tourist is a traveller who takes a trip to a destination outside of his or her normal environment for more than a day and less than a year. Travelling for sightseeing is one of the common understandings of tourism and wildlife tourism can be a good example.

While defining wildlife tourism, views of several wildlife tourism experts are crucial. Higginbottom and Tribe (2004) define wildlife tourism as tourist activity having wildlife as the main focus. On the other hand, Beeton (2004) argues that the scale and size of wildlife tourism enterprises can considerably differ, ranging from mass tourism oriented aquaria and large zoos to specialized wildlife tourist-attracting private parties.

5.2.1 WILDLIFE TOURISM VERSUS ECOTOURISM

Before defining wildlife tourism, it is particularly important to make a clear distinction between ecotourism and wildlife tourism. This is important because both of these terms very often interchange with each other, creating confusions. According to The International Ecotourism Society (2015), ecotourism is "responsible travel to natural areas that conserve the environment and improve the well-being of local people." In particular, ecotourism concentrates on ecological experiences in its natural environment. Even though the basic goal of ecotourism is to relate to nature, wildlife species are an unavoidable part of a comprehensive ecological setting. Thus, this is common that all tourism activities in natural areas cannot be defined as ecotourism. Still, the interrelationship between ecotourism and wildlife tourism remains strong. In ecotourism, the natural environment needs to be safeguarded against negatives to bring more positive impacts. Awareness creation among local populations regarding environmental and wildlife sensitivity becomes important. Financial supports to improve ecological and wildlife situations are important as well for both assuring and empowering local communities. These are particularly true for both ecotourism and wildlife tourism (Saleh & Karwacki, 1996; Weaver et al., 1996).

5.2.2 WILDLIFE TOURISM VERSUS WILDLIFE-WATCHING TOURISM

Wildlife tourism involves all scales and types of tourism for enjoying the wildlife and natural scenic beauty. Wildlife tourism can be outlined as a particular tourism type that refers to both non-consumptive and consumptive viewing of wildlife species and resources in natural settings (Newsome et al., 2005). This form of tourism has either high or low impacts from the humans, is capable to generate low or high economic outcomes, can be unsustainable or sustainable, international or domestic, and can deal with both longer stays and short day visits (Roe et al., 1997). Again, some sort of distinctiveness exists between this type of tourism and wildlife-watching tourism. The latter is merely an activity that involves the watching of wildlife species (STCRC, 2009). This prototype of wildlife tourism generally refers to watch species differentiating with general wildlife tourism that is much wider in scope and ranges from watching to activities such as fishing and hunting. Wildlife-watching tourism is importantly featured as an observational activity, which can involve interactions with animals such as watching, feeding, and touching them.

Wildlife-watching tourism, in general, is undertaken or organized for watching wildlife in its natural habitat. Such a type of tourism has managed rapid popularity. However, in common, the tourism industry mostly uses the term "wildlife tourism" than "wildlife-watching tourism;" these terns have differences between them. Visibly, these two terms appear similar; these are identical and differentiated to each other. Wildlife tourism in certain cases is referred to fishing or hunting or simply viewing captive wildlife in confined parks or zoos. Wildlife-watching tourism is meant to include watching on big game ranches as in Southern Africa where diverse species can roam freely in relatively bigger ranges (Christie et al., 2013). Also, their management becomes critical when their behavior becomes naturally wild.

5.2.3 WILDLIFE TOURISM VERSUS EDUCATION TOURISM

The term "Education tourism" or "Edu-tourism" refers to any "program in which participants travel to a location as a group with the primary purpose of engaging in a learning experience directly related to the location" (Bodger, 1998, p 28). This type of tourism may be categorized into the following dimensions: cultural/historical, eco-tourism/nature-based tourism/rural tourism, and study abroad programs (Ankomah & Larson, 2004). "Educational tourism" is a tourist activity undertaken by those who are undertaking an overnight vacation and those who are undertaking an excursion for whom education and learning is a primary or secondary part of their trip (Ritchie, 2003, p 18). It comprises several sub-types including ecotourism, heritage tourism, rural/farm tourism, and student exchanges between educational institutions. The notion of travelling for educational purposes is not new (Gibson, 1998; Holdnak & Holland, 1996; Kalinowski & Weiler, 1992), while its popularity in the tourism market increases as per expectation (Gibson, 1998; Holdnak & Holland, 1996). Lifelong learning (LLM) is "all learning activity undertaken throughout life, with the aim of improving knowledge, skills and competence, within a personal, civic, social and/or employment-related perspective" (Europa, 2003). Education tourism can play a vital role in learning process of the knowledge seekers. Rather than directly visiting a wildlife sanctuary it will be beneficial if the wildlife lovers/watchers first have information through tutorials. It will create a connection between the wildlife watchers and animals. Secondly, it will also create awareness about the wildlife habitat to the visitors (Fig. 5.1).

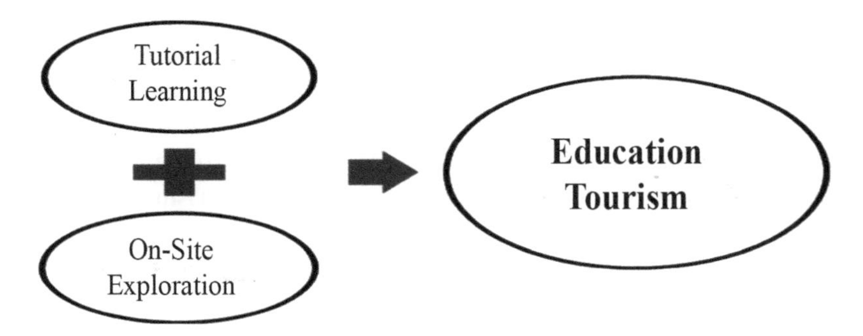

FIGURE 5.1 Education tourism model (*source*: the authors).

5.3 WILDLIFE TOURISM MARKETING

In countries across the world, the contributions of wildlife tourism are increased significantly and wildlife tourism is becoming global by eliminating geographical barriers. Thus, marketing activities for wildlife tourism are becoming diverse with added importance of tourism promotion in countries across the world.

According to Diaz-Perez et al. (2005) and Scottish Government Social Research (2014), the profile of wildlife tourists is diverse of many features and characteristics. In general, wildlife tourists tend to stay short in more places and keep on moving from one place to another, which is, an unusual nature of general tourists. Wildlife tourists aspire to visit more places having wildlife species both locally and globally. These tourists tend to spend low for covering their daily expenses, but can spend higher mainly due to their longer stays in different places. The age group of wildlife tourists is likely to be on the lower side and they mostly travel in a group of friends or couples. Origin of this tourist segment is also diverse, ranging many countries of the world including African, North American, Asian, and European countries. Wildlife tourists mainly make their visit as their first and only visit, as part of the business, holiday or visiting family, or friend's purposes. Wildlife tourists are relatively keen to visit animals as shark, crocodiles, parrot, snake, lizard, penguin, turtle, seal, and fish. Again, visitors on their first trip are keen to see iconic marsupials including koala or kangaroo and visitors on a return visit are eager to see any type of animals that they have not seen during their last visit. In common, the satisfaction level and engagement of these tourists with wildlife species are relatively higher than in other tourism types. By nature, people's longevity and affluent living in industrial societies enable them to become closer with nature and the environment. This becomes more regular during the time they retire from services or jobs and they tend to look for new experiences through participating tourism activities and wildlife tourism is an example of this case.

The growth of travel and tourism has been remaining enormous over the last few decades. According to Tapper (2006), the number of international tourist arrivals was 441 million in 1990 growing to 763 million in 2004 with 52% of these arrivals for leisure and recreational tourism activities. The pace of growth expects to continue at an estimated 1.6 billion international arrivals in 2020. Also, domestic tourism is increasing around the world followed by people's escalating capacities of spending of both leisure time and money to participate in tourism activities. The measurement of domestic tourism situation across countries is a bit difficult due to diversities in measuring or indexing indicators. However, in general, the estimation is that domestic tourist number is 10 times higher than international tourist arrivals in a country that has faster growth in coming years. Wildlife tourism also shows equal growth prediction. One of the basic reasons is the commercialization of wildlife tourism activities. This type of tourism is getting more closely interrelated with marketing that creates enormous interests among diverse groups of tourism entrepreneurs and general tourists. More tour operators and travel agents are concentrating on tourism, ensuring sustainability and environment-friendly products and services. These marketers are also relatively keen to offer wildlife-friendly products and services ensuring carbon-neutrality and local people's participation, in terms of benefit and livelihood generation.

According to Robinson (2012), wildlife tourism has been playing a crucial role in strengthening economic activities of specific countries including Seychelles, East Africa, or the Galapagos Islands. This type of tourism has been acting as the base of their national economy. More specifically, wildlife-watching tourism as part of wildlife tourism is considered as a relatively newer form to help diversify tourism products and services to attract tourists. This also helps to promote community development activities in remote areas of these countries. For example, the number of whale watching tourists was nearly 4 million in 1991 that has risen to 9 million in 1998 where, the total revenue from whale watching stands at over US$

1 billion as three times higher from 1991. This revenue generation helped at least 495 communities around the world to become able to develop their livelihood and living standard. During 2003 and 2004, a research showed that the number of whale watchers in and around Sydney has doubled and the revenue generation increased to four times. The rate of such growth continues with the increasing number of international tourists. Still in Australia, some areas are relatively more popular than the others in relation to wildlife watching. Sydney is considered as relatively more popular. The key reasons for the tourist number rise are diverse factors including long-term interests in wildlife species and their availability.

5.3.1 AN EXAMPLE

Marketing for wildlife tourism becomes more dominant which is evidenced from the Cayman Islands. Major tourist attractions in this island including the Sand Bar and Stingray City have unique marketing strategies. Wildlife tourism marketing activities actually relates the entire island. These two places in the Cayman Islands are located in shallow waters of North Sound in Grand Cayman. Both these sites are famous for diving and snorkeling with stingrays. In general, a minimum of 30 stingrays can be seen at Stingray City and 50 stingrays at the Sand Bar. One of the rare features of swimming with stingrays is the opportunity to both feed and touch these animals those are relatively rare in these days. In these places, around 900,000 visits are made on average each year. Out of these, around 780,000 visits are made by cruise passengers. These visits are commonly made to the Stingray City and Sand Bar meaning that half of the visitors coming to the Cayman Islands take this particular trip. The Sand Bar is 60 cm deep in many places allowing snorkelers to shallow and become able to touch stingrays resting on the bottom. On the other side, the Stingray City is relatively deeper with 3–5 m depth. This place is mostly visited by scuba divers coming for recreation. For the Cayman Islanders, the stingray experience by tourists is mostly important that also attract a huge number of tourists to this island. This activity allows tourists to contribute to this island's economy. Almost one-fourth of the island's economy heavily relies on tourism contributing to the local economy (United Nations Environment Programme and Convention on the Conservation of Migratory Species of Wild Animals, 2006).

5.3.2 *ANOTHER EXAMPLE*

Australia is a famous wildlife tourism destination in the world. A recent survey about wildlife tourism in Australia shows that the satisfaction level of tourists is generally calculated by matching their interests to see specific animals during their visit and the availability of those specific wildlife species. Following such calculation, most of the animals matched to the interests of tourists meaning a success rate over 50%. In particular, the parrot has the highest matching rate with 81.8%, the Kangaroo with 81.1%, and the Koala with 74.7%. The lowest matching rate is for the Platypus with 44.3% and the Whale with 17.1%. In general, 81.4% of the total tourists in the sample were satisfied with their wildlife experiences and 98.4% indicated that they were satisfied with their overall visit to Australia (Sustainable Tourism Cooperative Research Centre, 2009).

5.4 TECHNOLOGY APPLICATION IN WILDLIFE TOURISM MARKETING

Tourism industry characteristically is sensitive to market demand and remains as a subject of influence by diverse factors those dominate a market structure. This is thus important that tourist demands are met by adopting specific technologies for marketing. Wildlife tourism in many countries is being commercialized rapidly and this becomes evident by increased wildlife visitations. Wildlife-watching opportunities of diverse species in different environments are becoming more popular and more realistic. The commercial aspects of wildlife tourism mainly cover special activities. However, wildlife tourism in most cases requires awareness about the environment and increasing capacities to interpret wildlife species. Technological gadgets are positively contributing tourism marketing (Dadwal & Hassan, 2015). Social networks mainly as Facebook and Twitter are playing crucial roles in promoting wildlife tourism in many countries across the world. Technology becomes finely tuned with wildlife tourism marketing.

5.5 WILDLIFE TOURISM CONSERVATION

The conservation of wildlife species for tourism activities is important. Rhinoceroses, elephants, tigers, or even great apes are mostly seen as iconic species for promoting wildlife tourism. But, illicit trading of these animals' parts is becoming highly profitable and lucrative. However, this is obviously an environmental crime. Demand for such animal parts is huge in the East Asian countries for medicine production or international commercial trading purpose. Also, exotic pets are very often seen as a symbol of status. According to the United Nations Environment Programme (2014) till date, illegal wildlife trade estimates to reach the worth of US$ 50–150 billion per annum and also, illegal fisheries catch around the world has reached to an estimated value of US$ 10–23.5 billion per annum. Elephants are one of the most targeted wildlife species for illegal trade meaning that illegal ivory trade as an example has doubled till date in less than seven years (Lawson & Vines, 2014). According to Nellemann et al. (2014) in Africa, at least 25,000 elephants are killed every year from a gross elephant population of 420,000–650,000. Ivory provides income to groups, such as the Lord's Resistance Army, currently operating in South Sudan, Central African Republic, and the Democratic Republic of Congo. The militia or insurgent groups in Africa tend to grab profits from illegal wildlife sales that actually funds terrorist activities (Wyler & Sheikh, 2013; Nellemann et al., 2014). Such illegal wildlife species trade also results in cartels and crime syndicates for illegal arms trading, money looting, or robbery. Rural communities are very often threatened by poachers to interrupt organized initiatives against them. These poachers obviously undermine the development and conservation of wildlife tourism in a country. Activities of such poachers actually drain the natural and ecological heritage of the regions destroying the entire wildlife that could possibly have been a reliable source of revenue generation. Activities of these poachers force African countries losing opportunities for tourism revenue and employment generation those could significantly contribute both the local and national economy. In general, due to such illegal wildlife trading and unexpected activities of poachers, some specific wildlife species are pushed 1000 times faster toward extinction than the desired natural rate (Pimms et al., 2014). The general situation thus shows a bleak picture of threatened wildlife tourism in parts of Africa, India, and several other natural resourceful countries of the world. This necessitates the demand to conserve, protect, and safeguard wildlife resources from the illegal trade, and dreadful activities of poachers. Conservation activities have greater potential for success if local people are allowed to take part in formulating and implementing policies and programs that incorporate safeguards against abuses and that place a strong emphasis on equity and social justice (Hitchcock, 1997).

5.5.1 THE EXAMPLE

According to The Guardian (2009) in the Cayman Islands, increasing concerns are rising regarding snorkeling and diving effects. The stingrays use to come to the Stingray City and the Sand Bar for feeding. Fish wastes mainly thrown by the overboard local fishermen are their food. However from 1986, these stingers are fed by visitors. The increasing presence of visitors has made changes in stingers' living style. In general, the stringers tend to move solitary but, these are recently forming groups of about 12–15 individuals. Also, these have changed behavior from nighttime feeding to daytime feeding in both places. Also, most of their foods come from divers rather, the local fishermen. A recent research made in conjunction to the Department of Environment shows that, the stingrays are becoming prone to injuries and exhibit a higher number of injuries resulted from increasing number of parasites on their grills, boat collisions, and open wounds. In relation to this, blood samples show that the stringers those are human fed are getting low-essential fatty acids. These are essential for immune response and disease resistance. Even the Cayman Islands have implemented a protected area networks, marine parks, and off coast zones. Still, the Stingray City and the Sand Bar remain outside of these networks and uncovered by protected area regulations. Since access to both of these areas is open, there is a visible threat of overcrowding resulting increased pressure on the stingrays. In order to address these issues more responsibly, the Cayman Islands Department of Environment has introduced a process aimed to attach stakeholder representatives from all parties including the public and private and the Marine Conservation Board to leave responsible roles (United Nations Environment Programme and Convention on the Conservation of Migratory Species of Wild Animals, 2006).

5.6 TECHNOLOGY APPLICATION IN WILDLIFE TOURISM CONSERVATION

In practice, technology application in wildlife tourism has been able to alter illegal trading of wildlife species including rhino horn, ivory, exotic birds, live apes, or pangolin scales. According to Lavorgna (2014), use of the Internet has actually enhanced wildlife trade and accelerated trafficking. Technology application has converted the entire world as a common market place having no barriers or distances. From the poacher user perspective, technology is also benefitting them. Through technology application on the scale of the Internet use, criminals can exchange information as faster and efficiently than ever before helping to expand their trading operations. Technology has also facilitated both exchange and communication of information and effective interrelationship between buyers, intermediaries, and suppliers to encourage illicit trading of wildlife species. Still, technology application plays definite and positive roles to support wildlife conservation, environmental protection and enforcing law, and order by relevant agencies. This is particularly evidenced from initiatives taken by many countries across the world including China.

5.6.1 THE EXAMPLES

Chian Government Network (2013) shows that, China launches the "Skynet Action" in 2013. This initiative is aimed to prevent wildlife trading and crimes with the support of e-commerce companies and websites.

In specific, the UNEP Global Environmental Alert report confirms that unmanned aerial vehicles (UAVs) or drones are brought into use in many parts of the world aimed to conserve wildlife (Platt, 2012; United Nations Environment Programme, 2013). According to Ol Pejeta Conservancy (2014), one of the pioneer countries using drones is Kenya. In June 2014, the Ol Pejeta Conservancy planned to deploy UAVs to keep monitoring illegal wildlife trading activities to protect mainly the rhinoceroses and elephants along with many other threatened wildlife species in the Mount Kenya area. This conservancy is one of the largest sanctuaries in East Africa to rear the black rhinos those are mostly seen as endangered species. This is a reason that continually attracts poachers for illicit activities. By the end of 2013, Kenya had a total 1041 rhino population as the third largest of that time. However by the first quarter of 2014, Kenya lost 59 and then 18 more rhinos by poachers (Kenya Wildlife Service, 2014).

As found, the use of the UAVs or drones as helpful to control illegal wildlife activities in the country. Still, the Kenyan government refused to grant approval and unexpectedly banned using private drones for "security risks" (Daily Nation Newspaper, 2014). In practice, anti-poaching operation has been slowed down following the ban to use drones for surveillance to monitor illicit poaching activities. On the contrary, this offered more freedom for poachers to perform their activities. This action by the Kenyan government is taken as an underscored action to challenge and eliminate illegal wildlife trading. In both practice and actions, an advanced technological tool as drones has been remaining as a key element for anti-poaching activities to accelerate and promote wildlife conservation.

Other than drones, law and wildlife enforcement agencies have started experimenting alternative technologies. The "Echo" technology or the acoustic traps are one of these alternative technologies. This technology monitors sound waves of sharp disturbances like chain saws, gunshots, blasts, truck engines, or aeroplane engines (Wrege et al., 2010; 2012). This technology is commonly used as conventional anti-poaching patrolling (Fig. 5.2).

5.6.2 TECHNOLOGY BASE IN WILDLIFE TOURISM

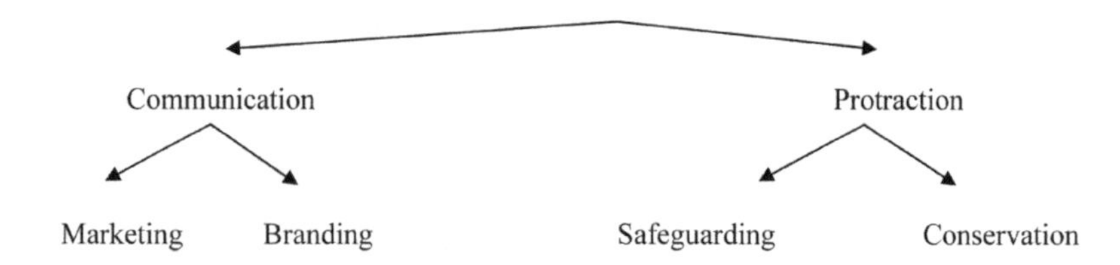

FIGURE 5.2 Technology base in wildlife tourism (*source*: the authors).

5.7 KEY CHALLENGES IN TECHNOLOGY APPLICATION IN WILDLIFE TOURISM

5.7.1 MARKETING

Wildlife tourism marketing is not evenly supported by both the public and private agencies across the world. This type of tourism denotes differences in terms of marketing policy formulations and implementations. One of the key challenges of wildlife tourism marketing is the lack of resources in countries in the world. The availability and distribution of resources are uneven in countries promoting wildlife tourism. This is a key aspect to consider effective wildlife marketing strategies to ensure proper use of resources and opportunities.

Technology use requires expertise and particularly for marketing. Thus, the creation of effective and workable technology experts still remains as one of the key challenges for marketing wildlife tourism. Marketing is a constant going process requiring monitoring and sound knowledge. The use of technology can help in this regard. This also becomes a challenge that wildlife tourism marketing necessitates academic knowledge about marketing as essential to successfully apply a certain technology to meet a desired purpose. This is a challenge also involves working or practical skills to implement specific technologies for marketing. Each county's academic structure and the system vary meaning that technology use for wildlife tourism marketing requires proper and adequate curriculum support to enrich knowledge. The availability of resources is also critical because technology use requires resource. Financial capacities of the involved parties need to be taken into consideration. The allocation and capacities to spend for technology use also require proper attention. This can be termed as a key challenge to use technology for wildlife tourism marketing.

5.7.2 CONSERVATION

The basic challenge of technology use for wildlife conservation in protected areas should be a commitment of the government (Eagles et al., 2002). This is clearly evidenced from the examples of Kenya as discussed in earlier part of this chapter. The government in many cases and many countries can refuse to use technology. Technology use can be successful to monitor, track, and eliminate illicit wildlife trade activities. This is also true that governments in many countries also have a limitation to allow wider and free use of technology for the sake of national security. The government has to consider the security of both the national and local populations, while the use of appropriate technology is also important. This becomes a basic challenge for technology use for wildlife conservation.

The other challenge is to make effective technologies as available for use. Technologies those are believed as effective for wildlife conservation tends to be relatively expensive. This is mainly the government's responsibility to make these technologies available for use mainly by public sector users (Higginbottom & Tribe, 2004). This is obvious that resources of each country cannot be same relying on many constraints as financial.

Developing countries can hardly use specific technologies in comparison to developed countries to conserve wildlife resources. This becomes a basic challenge for technology use to conserve wildlife tourism.

The other challenge regarding technology use for wildlife tourism conservation is the strong financial, armed, and other resource capacities of poachers. These poachers in many cases pose huge resources to bring in use any sophisticated technology to tackle and down the available technologies as mostly used by government agencies. This is a practical challenge to make updated technologies unavailable and inaccessible for use by the poachers to defend illicit activities and conserving wildlife tourism.

5.8 CASE STUDIES

5.8.1 UNLOCKING POTENTIALS OF THE SUNDARBANS WEST WILDLIFE SANCTUARY, BANGLADESH

The Sundarbans West Wildlife Sanctuary is an animal sanctuary in Bangladesh and also a UNESCO World Heritage Site. Geographically, this is a key part of the wider the Sundarbans region. The Sundarbans is the world's largest mangrove forest. The forest is a joined delta of the Brahmaputra, Meghna, and Ganges rivers having close proximity of the Bay of Bengal. The Sundarbans has an approximate coverage of a million hector of land area. Among this, 60% land area remains to Bangladesh and the rest to India, while the region is divided by the Raimangal River (Fig. 5.3).

FIGURE 5.3 Wildlife tourists inside the Sundarbans (*source*: Deshghuri, 2015).

The Sundarbans within the Bangladeshi geographical area includes three wildlife sanctuaries as the Sundarbans South, the Sundarbans East, and the Sundarbans West. The Sundarbans West Wildlife Sanctuary has a total of 715 square kilometers as covered by this sanctuary. In general, the region is so many complexes have intersected by a network of small islands, mud flats, and tidal waterways for nurturing salt tolerant mangrove forests. As a common feature, the area is mostly flooded with brackish water during the time of high tides mixing freshwater from inland rivers. This feature allows the Sundarbans to have several mangroves including the hantal palm (*Phoenix paludosa*), the Gewa (*Excoecaria agallocha*), the Goran (*Ceriops tagal*), and many others. On the other side, the fauna of this sanctuary is also diverse with at least 260 bird species, 35 reptiles, and 40 species of mammals. Also, at least five species of marine turtles with two nearly extinct reptiles as the Indian python and the estuarine crocodile are seen in the coastal zone. The Chital Horin (spotted dear), macaque monkey, the Indian otter, and the wild boar are some other wild-life species. Out of all these, the Royal Bengal Tiger is the most prominent. However, the number is gradually decreasing as this becomes a target of the poachers (UNESCO, 2015).

For marketing, the Sundarbans West Wildlife Sanctuary has exploited potentials that can be simply immense. However, marketing approaches and strategies are less visible in this sanctuary creating more interests of concerned parties and interest groups. The responsible authority for this sanctuary is Bangladesh Forest Department that as a governmental agency visibly has fewer roles for marketing and to make this sanctuary as known to global wildlife tourists segments. Existing national tourism policies are favorable for tourism promotion in Bangladesh (Hassan & Burns, 2014). Tourism marketing potentials are not fully trapped requiring more involved of environmental and interest groups with promotional initia-tives (Hassan, 2014; Hassan, 2012b). On the other side, private wildlife tour operators are also becoming more interested to organize tours in this sanctuary. Technology is in use for marketing this sanctuary where, social networking sites are playing important role for marketing of this wildlife sanctuary. As well as websites of the involved authorities as the Ministry of Forest and Environment of the Government of Bangladesh, Bangladesh Forest Department, and importantly, the UNESCO are playing role for its marketing. Online reviews from visitors also help.

Regarding conservation, the sanctuary belongs to the UNESCO World Heritage Site. A UNESCO WHS gets considerable attention from international agencies (Hassan & Iankova, 2012). However, illegal activities can never be stopped even; activities of the concerned agencies are mentionable in this regard. The Royal Bengal Tiger and the Chital Horin (spotted dear) are the two main wildlife species. The recent incident of oil spillage inside the Sundarbans manages more attention from concerned authorities and parties. The oil spillage incident from a sunken oil tanker with 350,001 of furnace oil in the Shela River on the December 9, 2014 has shaken the entire wildlife tourist community. The incident severely threatened the mangroves with endangered wildlife species mainly the Ganges and Irrawaddy dolphins. However, the incident brings the resource limitation of the country to response a calamity. The availability of technologies is so inadequate that local peoples are involved for oil collection. International agencies at the later stage were invited to play a role in collection oil and minimize the harmful effects of oil spillage.

The Sundarbans West Wildlife Sanctuary covers pristine and purely natural settings. Being less intervened by technologies, the sanctuary can offer exclusive wildlife experiences. There are true potentials for marketing and relevant agencies need to come forward to capitalize these unexplored potentials to the maximum. On the other side, the sanctuary is less likely to be heavily threatened by illicit and illegal poaching activities. Government agencies are playing mentionable roles to both carry out and support conservation activities to a desired level. Still, there are scopes for the betterment of both marketing and conservation with appropriate technology application.

5.8.2 EXPLORING THE UNEXPLORED WILDLIFE TOURISM

The Sorsan Wildlife Sanctuary, Hadoti Region, Rajasthan, India: The Sorsan wildlife sanctuary is located near Baran District in the Hadoti region of Rajasthan, India. Around 1800 black bucks, 324 chinkaras, Indian fox, monitor lizard, and over 100 species of birds like the chestnut-bellied, black-winged kite juvenile, kristel, and many more are visible in this sanctuary. The sanctuary can be developed not only to view animals and birds but also more offers for tourists. The Sorsan Blackbucks Sanctuary has high tourism potential followed by good transportation network from

Kota. Comfortable accommodation options with good amenities are available in nearby cities as Anta, Kota, and Baran with an average distance of 20, 65, and 45 km, respectively. The Amalsara is a nearby place to the sanctuary where six big and beautiful huts are constructed with an investment of 5 million Indian Rupees to accommodate tourists. The Sorsan is a rare place where one can experience spirituality coupled with wildlife adventures and rural lifestyles. The Brahmani Mata Temple, the Amalsara Huts, and the Nagda—all three religiously sacred establishments conjoin this sanctuary to allow tourists to stay and experience spirituality within a closer vicinity of the natural environment.

Tourism potentials in the Sorsan Blackbucks Sanctuary to attract passionate tourists to see wildlife species in natural habitat and in rural lifestyle are immense. There is an urgency to take effective marketing strategies. Marketing becomes important for the promotion of this sanctuary as a wildlife tourism destination. Expected spending from increased tourists may be more, while contributions from local peoples are regular and reliable. Thus in present situations, both the government and wildlife tour operators need to focus on programs aiming to be able to attract more tourists. The sanctuary itself and nearby places lack many required factors as organizational initiatives, growth of a mutual path of benefits in synergies, market competitions (competition between the Sorsan and other sanctuaries of Hadoti), promotional or branding activities, and even a logo.

Also in the digital era, the sanctuary is not at all technology connected. This is shocking that there is no website of the sanctuary and nearby visiting places and thus, destination marketing approaches are missing. For marketing, technology use for the development of a website with full of information and attractions is much required. A virtual tour should be designed on online platforms including photographs and videos with traditional or folklore music. The Sorsan Wildlife Sanctuary website can also include temple and black bucks tour, village walk tour, spirituality tour, video library with wildlife tourism aspects, information on the weekly market/haat bazaar, and wildlife movies. Added emphasis should be placed to use and capitalize m-commerce, e-commerce, social media, blogging, and so on. Other than using updated digital technological approaches, traditional methods need to be used for the promotion like leaflets, books, press coverage, and related. Local peoples at large can be involved with market promotional activities. Tourist demands need to be addressed by analyzing tourist needs and feedback. Visitors are required to be communicated by using a variety of media.

For facility enhancement, nature green park café for breakfast, lunch and dinner can be established at Amalsara (with traditional Hadoti Dishes and Organic Food). Even facilities as Spa, Mud Bath, Yoga, and Meditation can also be planned to implement. The huts of Amalsara can be used as Adventure Forest Resort with facilities as trails, safaris, traditional dances, and so on. Open membership scheme for the resort and sanctuary can be introduced. Employing skilled workers from the remote and rural area and to train or retain them for service offers. Stable funding sources with networking is required and starting wildlife viewing to allow visitors, tourists, and local people understand the importance of the sanctuary to minimize negative impacts on wildlife species. Information sharing followed by direction guidance as trails, signs, brochures, or maps can encourage wildlife tourism.

It is very well identified the Sorsan sanctuary has an immense opportunity for wildlife tourism promotion and to attract tourists from across the world. However, any mentionable initiative for conservation is missing. The sanctuary is not also associated with many other types of tourism activities to ensure conservation initiatives. At present, there has not been any seen or reported an incident of threats to wildlife species. Still, the basic need of the sanctuary is to prepare and implement sound strategies for its conservation. This is the high time when the government and local population have to come together for the conservation of the Sorsan sanctuary. Tourists started to flock to the Hadoti region with the purpose of wildlife tourism meaning that conversation strategies are required (Fig. 5.4).

For conservation, creating a network for effective technology application between biologists, researchers, staff, and other peoples can help. Strictly, there should not be touching, feeding, and taking photos using flashes with wildlife species. The government efforts for the conservation of wildlife tourism by using appropriate technology become important. For the survival and taking the Sorsan wildlife sanctuary on the right path, steps are required to be taken. For conservation, the government, as well as the researchers, needs to be getting involved. The government needs to encourage technology use to support wildlife tourism in the region. Conservation activities need to actively involve research and monitoring of target wildlife species. Research and support are also needed to facilitate economically feasible development initiatives of wildlife tourism in private land areas to assist conservation. There is a need to determine under what circumstances, and by what means, landowners can be financially benefitted from such shift. Similarly, an examination of existing mechanisms and constraints on such shifts should be conducted. Financial incentives should be offered to land owners who switch from relatively unhelpful land-uses to well-managed wildlife tourism. Research is also needed to assess the effects of wildlife tourism on public attitudes toward conservation, and how benefits can be maximized. Research is required to determine the potential role of local communities in participation and support for integrated wildlife tourism conservation initiatives followed by the increasing role of social and economic incentives. In addition, research is also wanted to determine the best possible ways to positively influence visitor attitudes and behavior regarding wildlife conservation through various forms of wildlife tourism activities. Government, conservation NGOs and tourism industry bodies should work together to strategically develop mechanisms for enhancing links between wildlife tourism and conservation. Mechanisms for encouraging tourists to make donations for conservation should be further developed. The sanctuary demands effective role play of both the private and public sector agencies. For fostering the growth and the wildlife tourism development the government as well as the local people have to take initiatives.

FIGURE 5.4 The Sorsan Wildlife Sanctuary, Rajasthan, India (*source*: the authors).

5.9 CONCLUSION AND IMPLICATIONS

The basic aim of this chapter is to outline the current status and potentials of technology adoption for wildlife tourism. The chapter cites a considerable number of examples covering parts of the world as important for wildlife tourism. Technology adoption for both marketing and conservation remains the central theme. Summarizing global examples, the chapter at the later stage explores untapped potentials of two relatively unknown wildlife tourism destinations of Bangladesh and India. Main concentration in the chapter is placed on marketing and conservation of wildlife species. This chapter agrees that the use of appropriate technology can be a boon for both marketing and conserving wildlife species and tourism activities from them. Emphasis needs to be placed on capitalizing the unlocked potentials of marketing and conservation. The study identifies the Internet as the reliable mean for marketing, while unique technologies need to be applied for conservation and reduce poaching or illicit wildlife activities. The study clearly defines that technology application can meet increasing demands of wildlife tourism. The potentials of emerging technologies are also enormous to use more acceptable technologies for a similar purpose. Many countries in the world including countries in Africa and Asia are using technology for marketing where, the chapter clearly cites examples of Kenya and China in this regard. The chapter stresses on active role play of the government allowing the use of innovative technologies. The recent development of technologies has in fact accelerated the process of technology application than ever before creating more space of usability and potentials for further use. Wildlife is an asset of the nature requiring responsibilities for both marketing and conservation, while its importance for both the education tourism and LLM becomes obvious. The chapter is evidenced that technology application can be of effective use for general wellbeing of wildlife species as well ensuring free movements of both wildlife tourists and wildlife species. This chapter as based on conceptual research mainly relies on available literature both printed and on the Internet. Still, empirical evidence could have expanded competencies of this chapter, that is, a limitation. Future research can encompass research on technology application containing empirical data and information.

KEYWORDS

- **technology**
- **wildlife**
- **tourism**
- **marketing**
- **conservation**

REFERENCES

Ankomah, P. K.; Larson, R. T. Education Tourism: A Strategy to Sustainable Tourism Development in Sub-Saharan Africa, 2004. Retrieved from: http://bit.ly/1J9aDYT

Azim, R.; Hassan, A. Impact Analysis of Wireless and Mobile Technology on Business Management Strategies. *J. Inform. Knowledge Manag.* **2013,** *2* (2), 141–150.

Beeton, S. Business Issues in Wildlife Tourism. In *Wildlife Tourism: Impacts, Management and Planning;* Higginbottom, K., Ed.; Common Ground Publishing: Victoria, US, 2004; pp 187–208.

Bodger, D. Leisure, Learning, and Travel. *J. Phys. Educ. Recreat. Dance.* **1998,** *69* (4), 28–31.

China Government Network (CGN), 2013. Forest Police Start "Skynet Action" to Protect Wildlife Resources. China Government Network (CGN). http://www.gov.cn/gzdt/2013-04/25/content_2389801

Christie, I.; Fernandes, E.; Messerli, H.; Twining-Wardet, L. Tourism in Africa: Harnessing Tourism for Growth and Improved Livelihoods, 2013. Retrieved from: http://bit.ly/1MsPXh2

Dadwal, S.; Hassan, A. The Augmented Reality Marketing: A Merger of Marketing and Technology in Tourism. In *Emerging Innovative Marketing Strategies in the Tourism Industry;* Ray, N., Ed.; IGI Global: Hershey, PA, 2015; pp 78–96.

Daily Nation Newspaper. *Government Bans Drone Use to Fight Poaching in Ol Pejeta,* 2014. Retrieved from: http://bit.ly/1tlyVbS

Deshghuri. Sundarban Tour Package, 2015. Retrieved from: http://bit.ly/1B1bd96

Diaz-Perez, F. M.; Bethencourt-Cejas, M.; Alvarez-Gonzalez, J. A. The Segmentation of Canary Island Tourism Markets by Expenditure: Implications for Tourism Policy. *Tourism Manage.* **2005,** *26* (6), 961–964.

Eagles, P.; McCool S.; Hains, C. D. *Sustainable Tourism in Protected Areas: Guidelines for Planning and Management;* The International Union for Conservation of Nature: Gland, Switzerland, 2002.

Europa. *European Commission: Policy Areas: Lifelong Learning, what is Lifelong Learning?* 2003. Retrieved from: http://bit.ly/1LcMrHa (accessed Aug 22, 2015).

Gibson, H. The Educational Tourist. *J. Phys. Educ., Recreat. Dance.* **1998,** *69* (4), 32–34.

Green, R.; Giese, M. Negative Effects of Wildlife Tourism on Wildlife. In *Wildlife Tourism: Impact, Management and Planning;* Higginbottom, K., Ed.; Common Ground Publishing: Victoria, US, 2004; pp 81–97.

Hassan, A. 'Package Eco-Tour' as Special Interest Tourism Product-Bangladesh Perspective. *Dev. Country Stud.* **2012,** *2* (1), 1–8.

Hassan, A. Key Components for an Effective Marketing Planning: A Conceptual Analysis. *Int. J. Manage. Dev. Stud.* **2012,** *2* (1), 68–70.

Hassan, A. Tour on an Imagined Heritage Trail Set in the Mosque City of Bagerhat, Bangladesh: Cogitation for Market Potentials. In *Tourism: Present and Future Perspectives;* Bansal, S. P., Walia, S., Rizwan, S. A., Eds.; Kanishka Publishers: New Delhi, India, 2014; pp 30–44.

Hassan, A.; Burns, P. Tourism Policies of Bangladesh – A Contextual Analysis. *Tourism Plan. & Dev.* **2014,** DOI: 10.1080/21568316.2013.874366.

Hassan, A.; Iankova, K. Strategies and Challenges of Tourist Facilities Management in the World Heritage Site: Case of the Maritime Greenwich, London. *Tour. Anal.* **2012,** *17* (6), 791–803.

Higginbottom, K.; Tribe, A. Contributions of Wildlife Tourism to Conservation. In *Wildlife Tourism: Impact, Management and Planning;* Higginbottom, K., Ed.; Common Ground Publishing: Victoria, Australia, 2004; pp 99–123.

Hitchcock, R. African Wildlife: Conservation and Conflict. In *Life and Death Matters: Human Rights and the Environment at the End of the Millenium;* Johnston, B., Ed.; Sage: London, 1997; pp 79–95.

Holdnak, A.; Holland, S. Edutourism: Vacationing to Learn. *Parks Rec.* **1996,** *31* (9), 72–75.

Kalinowski, K.; Weiler, B. Review. Educational Travel. In *Special Interest Tourism;* Hall, C. M., Weiler, B.; Eds.; Bellhaven: London, 1992; pp 15–26.

Kenya Wildlife Service 2014; Kenya Wildlife Service Statement on Status of Wildlife Conservation: Africa, 2014. Retrieved from: http://bit.ly/1B1bd96

Lavorgna, A. Wildlife Trafficking in the Internet Age. *Crime Sci.* **2014,** *3* (5), 2–12.

Lawson, K.; Vines, A. The Costs of Crime, Insecurity, and Institutional Erosion. 2014. Retrieved from: http://bit.ly/1JDqK0c

Nellemann, C.; Henriksen, R.; Raxter, P.; Ash, N.; Mrema, E. *The Environmental Crime Crisis - Threats to Sustainable Development from Illegal Exploitation and Trade in Wildlife and Forest Resources: A UNEP Rapid Response Assessment*; United Nations Environment Programme and GRID-Arendal: Nairobi and Arendal, 2014.

Newsome, D.; Dowling, R. K.; Moore, S. A. *Wildlife Tourism;* Channel View Publications: Toronto, Canada, 2005; p 16.

Ol Pejeta Conservancy 2014; Defining the Future of Drones in Conservation: Ol Pejeta Conservancy and Airware Test the Aerial Ranger TM in Kenya. 2014. Retrieved from: http://bit.ly/1hREHsf

Pimm, S.; Jenkins, C.; Abell, R.; Brooks, T.; Gittleman, J.; Joppa, L.; Raven, P.; Roberts, C.; Sexton, J. The Biodiversity of Species and their Rates of Extinction, Distribution, and Protection. *Science*. **2014,** *344*, 987–997.

Platt, J. *Eye in the Sky: Drones Help Conserve Sumatran Orangutans and Other Wildlife;* 2012. Retrieved from: http://bit.ly/1rMM63l

Ritchie, B. W. *Managing Educational Tourism;* Channel View Publications: Clevedon, UK, 2003; p 18.

Robinson, P. *Tourism: The Key Concepts*; Routledge: Oxon, UK, 2012.

Roe, D.; Leader-Williams, N.; Dalal-Clayton, D. *Take Only Photographs, Leave Only Footprints: The Environmental Impacts of Wildlife Tourism;* Environmental Planning Group: London, 1997.

Saleh, F.; Karwacki, J. Revisiting the Eco-tourist: The Case of Grasslands National Park. *J. Sustain. Tour.* **1996,** *4* (2), 61–80.

Scottish Government Social Research 2014; The Economic Impact of Wildlife Tourism in Scotland. 2014. Retrieved from: http://www.gov.scot/Resource/Doc/311951/0098489.pdf

Sustainable Tourism Cooperative Research Centre (STCRC) 2009; WILDLIFE Tourism: Challenges, Opportunities and Managing the Future. 2009. Retrieved from: http://bit.ly/1AXP242

Tapper, R. *Wildlife Watching and Tourism: A Study on the Benefits and Risks of a Fast Growing Tourism Activity and its Impacts on Species;* 2006. Retrieved from: http://bit.ly/1RYKCOm

The Guardian 2009; Stingrays Suffering from Contact with Wildlife Tourists, Study Finds. 2009. Retrieved from: http://www.theguardian.com/environment/2009/may/29/wildlife-tourism-stingray

The International Ecotourism Society 2015; What is Ecotourism? Retrieved from: http://bit.ly/1tlyVbS

The United Nations Educational, Scientific and Cultural Organization (UNESCO) 2015; Sundarban Wildlife Sanctuaries: Bangladesh. Retrieved from: http://bit.ly/1Gm37II

United Nations Environment Programme (UNEP) 2013; A New Eye in the Sky: Eco-Drones. Retrieved from: http://bit.ly/1Gm37II

United Nations Environment Programme (UNEP) 2014; UNEP Year Book 2014; Emerging Issues in Our Global Environment. United Nations Environment Programme: Nairobi, Kenya, 2014.

United Nations Environment Programme and Convention on the Conservation of Migratory Species of Wild Animals 2006; Wildlife Watching and Tourism: A Study on the Benefits and Risks of a Fast Growing Tourism Activity and its Impacts on Species. Retrieved from: http://bit.ly/1QiuIko

Weaver, D.; Glenn, C.; Rounds, R. Private Ecotourism Operations in Manitoba, Canada. *J. Sustain. Tour.* **1996,** *4* (3), 135–146.

World Tourism Organization (UNWTO) 2015; Understanding Tourism: Basic Glossary. 2015. Retrieved from: http://media.unwto.org/en/content/understanding-tourism-basic-glossary

World Travel and Tourism Council (WTTC) 2014; Travel and Tourism Economic Impact 2014 World. 2014. Retrieved from: http://www.wttc.org/-/media/files/reports/economic%20impact%20research/regional%20reports/world2014.pdf

Wrege, P.; Rowland, E.; Bout, N.; Doukaga, M. Opening a Larger Window onto Forest Elephant Ecology. *Afr. J. Ecol.* **2012,** *50*, 176–183.

Wrege, P.; Rowland, E.; Thompson, B.; Batruch, N. Use of Acoustic Tools to Reveal Otherwise Cryptic Responses of Forest Elephants to Oil Exploration. *Conserv. Biol.* **2010,** *24* (6), 1578–1585.

Wyler L.; Sheikh, P. *International Illegal Trade in Wildlife: Threats and U.S. Policy Congressional Research Services*. 2013. Retrieved from: http://bit.ly/1JDpLgz

CHAPTER 6

THE ANALYSIS OF TOURISTS' INVOLVEMENT IN REGARD TO DOLPHIN INTERACTIONS AT THE DOLPHIN DISCOVERY CENTRE, BUNBURY, WESTERN AUSTRALIA

LI CONG[1*], DIANE LEE[2], DAVID NEWSOME[3], and BIHU WU[4]

[1]*Tourism Management Department, Landscape School, Beijing Forestry University, Beijing, China*

[2]*Tourism and Events Program, School of Arts, Murdoch University, Perth, Western Australia*

[3]*Environment and Conservation Program, Murdoch University, School of Veterinary and Life Sciences, Perth, Western Australia*

[4]*International Centre for Recreation and Tourism Research, School of Urban and Environmental Sciences, Peking University, Beijing, China*

[*]*Corresponding author. E-mail: congli1980@163.com*

CONTENTS

ABSTRACT

Wildlife tourism is well developed and accepted in Australia, and is regarded as an opportunity for both wildlife conservation and regional economic development. This research aims to explore the involvement of wildlife tourists in Dolphin Discovery Centre (DDC) Bunbury, West Australia. A variety of empirical quantitative analysis methods, including the construction of a measurement scale for place involvement, K-means cluster, independent sample t-test, ANOVA, and Scheffe's post hoc test were used. The main findings for this research are taxonomies were proposed according to the cluster analysis: place involvement was divided into three categories according to its overall extent level with scale measurement: deep place involvement (DPI), medium place involvement, and shallow place involvement (SPI). It was found that nationality, age, and family background had significant difference on place involvement ($p < 0.05$), while gender, educational background did not ($p < 0.05$). Tourist behaviors, such as, length of stay, place consumption, source of information, travel companion, and satisfaction also had significant differences in involvement ($p < 0.05$).This research could provide the decision-making base for wildlife also. They could be applied to tourism destination marketing in DDC and might extend to other regions with similar situation.

> **"Non-consumptive wildlife tourism marks a clear shift from the traditional consumptive uses of wildlife resources."**

6.1 INTRODUCTION

Non-consumptive wildlife tourism is well developed and accepted worldwide, and is commonly regarded as an opportunity for the economic development and wildlife conservation (Curtin, 2010; Duffus & Dearden, 1990). It has been estimated that the economic value of wildlife tourism in Australia has been between $AUD1.8 and 3.5 billion. The majority of research on wildlife tourism concentrates in developed countries, while the academic attention in developing countries is seriously insufficient (Cong et al., 2014; Cong et al., 2015). Based on involvement theory, involvement is one of the important issues in the study of consumer behavior. The degree of tourists' involvement in destinations will affect their behavior and will also determine their satisfaction and revisit rate (Zaichkowsky, 1985). Exploration on the degree of wildlife tourists' involvement in destinations and its differences is meaningful to the understanding of purchase decision and behavior in destinations.

6.2 LITERATURE REVIEW

6.2.1 NON-CONSUMPTIVE WILDLIFE TOURISM

Duffus and Dearden (1990) defines non-consumptive wildlife tourism as "human recreational engagement with wildlife where the focal organism is not purposefully removed or permanently affected by the engagement." Non-consumptive uses of wildlife resources involve varied activities with a multiplicity of levels of organization all of which will influence the level and types of its impact (Boyle & Samson, 1985). Non-consumptive wildlife tourism marks a clear shift from the traditional consumptive uses of wildlife resources.

Early research on non-consumptive wildlife tourism focused on the influences of "Environment-Biology" systems of tourism (Sindiyo & Pertet, 1984). The focus of research was generally on a single product from several tourism types, such as nature-based tourism, ecotourism, and adventure tourism to observe and analyze the phenomenon of wildlife tourism. From a Chinese cultural perspective and after carefully reviewing foreign scholars' literature of wildlife tourism, Cong et al. (2012) and other scholars found that Chinese researchers in this field started late and involved multi-disciplines. The research achievements mainly concentrate on tourists' experience of visitors, the influence analysis of wildlife tourism, and destination development and management which establishes a core research framework, "tourists' experience-species conservation-destination development" (Cong et al., 2012).

In 2010, Catlin and Jones presented their results from a vertical analysis of whale watching in 1996–2006, Australia. They found whale watching is a tourism activity with adventure, with increasing tourism numbers and is gradually becoming a mainstream tourism activity with wide age and cultural distribution. They found that whilst most of the whale-watching tourists are Australian with a niche market, more and more foreigners, especially the Japanese, are willing to participate in the activity (Catlin & Jones, 2010). Research on wildlife tourism has concentrated developed countries, and perspectives from eastern countries, especially from China are relatively rare (Cong et al., 2012; Ma & Cheng, 2008; Catlin & Jones, 2010). Some Chinese scholars have begun to explore wildlife tourism from the perspective of supply, emphasizing the survey of current status of wildlife resources, presenting the existing regional wildlife tourism product, and providing specific measures for the development of tourism (You & Dai, 2010; Wang, 2008). Huang Zhenfang and other scholars have analyzed the spatial characteristics of eco-tourism flows and constructed models to predicate the number of tourists by various mathematical methods (Huang et al., 2007; Huang et al., 2008; Xu, 2004; Yuan et al., 2007; Yuan et al., 2009). In terms of a pioneer study of bird watching tourists in Beijing, Li (2009) used and empirical analysis of the demographic characteristics of bird watching tourists and found that young and middle-aged tourists are the main participants. The bird watching tourists were more highly educated, had slightly higher male tourists than female tourists, and were at the low to medium levels of income. Bird watching tourists could be divided into four categories, namely popular type, occasional type, positive type, and skilled type (Li, 2009).

Research on non-consumptive wildlife tourism has just begun with the current focus on wildlife tourism and its environmental influences in zoology and ecology (Catlin & Jones, 2010). At the applied level, there are many issues (Xu, 2004). One of the key gaps found in this review of the literature is that little research has been undertaken on the involvement of wildlife tourists in the wildlife site.

6.2.2 INVOLVEMENT

Involvement is one of the most important issues of consumer behavior (e.g., Hu & Yu, 2007), which is a significant dimension of understanding consumers' purchasing decisions (McGehee et al., 2003). Zaichkowsky (1994) classified the dimension into high involvement and low involvement. Early research, based on individual's behavior, Houston and Rothschild (1978) classified the notion of involvement into the following categories:

- situation involvement;
- enduring involvement;
- reaction involvement.

Laurent and Kapferer (1985) proposed a Consumer Involvement Profile (CIP) scale and obtained five antecedent variables affecting involvement:

1. importance,
2. entertainment,
3. symbol or symbolic value,
4. importance of risk,
5. possibility of risk.

Zaichkowsky (1994) developed the Personal Involvement Inventory (PII) scale based on a questionnaire with 20 items. This scale measured:

- advertising involvement,
- product involvement,
- purchase involvement.

The CIP scale is a multi-dimensional method and is more widely used because the PII scale is one-dimensional and its function is limited when measuring complex involvement. In order to measure research objects accurately, many scholars combine these two scales and modify them appropriately (e.g., Hwang et al., 2005; Carneiro & Crompton, 2010).

Kim et al. (1997) examined the relations between personal involvement and future travel wishes during bird watching tourism and found the measure involvement is more useful than the measure of social psychology. Their paper proposes five measure factors, namely behavior and membership system, the number of identifying bird species, behavior of bird watching in Texas, and consumption outside Texas (Kim et al., 1997).

In tourism, research on place involvement is still in its infancy. Gursoy and Gavcar (2003) examined Laurent and Kapferer's (1985) CIP scale in a Turkish international leisure tourism site and through the application of exploratory factor analysis, confirmatory factor analysis, and reliability and validity examination, a tourist involvement scale with three dimensions was confirmed. These involvement dimensions are as follow:

- entertainment/interest;
- possibility of risk;
- consequence of risk.

In contrast to the CIP scale, the symbol or symbolic value did not appear as a separate dimension but was loaded on entertainment/interest and possibility of risk. The three dimensions of Gursoy and Gavcar are consistent with results of previous studies (Havitz & Dimanche, 1999).

Although scholars have different opinions about the involvement dimensions, most of them agree that the concept of involvement has a multi-dimensional structure in which interest/importance is a well-recognized and significant dimension (Havitz & Dimanche, 1997). Prayag and Ryan (2012) used structural equation modeling to show that destination image, personal involvement, and attachment toward local places are the premise of tourists' loyalty. They also noted that tourist satisfaction could also have an effect.

Cai et al. (2004) examined the relationship between tourists' purchase decision involvement and information search behavior. In 2009, Carneiro and Crompton studied the influences that contact degree, structural constraint, and involvement have on tourists' search behavior when looking for destination information. They found that the relationship between involvement and information search behavior was provisional. Involvement is relatively stronger when tourists make early decision. Those who are unfamiliar to the destination tend to do more information search (Zhang & Lu, 2010). When Lehto et al. (2004) studied the tourists' revisit phenomenon, they took "consumer involvement" as the basic theory and described tourists' destination involvement as a two-dimension concept, namely behavioral dimension and psychological dimension, specifically including four dimensions: pre-travel involvement, risk involvement, activity involvement, and economic involvement. These four dimensions are considered as dependent variables involved and variables of psychologically related tourists and the outcome of destination involvement (Lehto et al., 2004). Destination involvement can be divided into four stages, respectively, destination awareness, destination attraction, destination attachment, and destination loyalty. Each stage is the gradual extension of psychological dimension (Beaton & Funk, 2008).

Hu (2003) examined the role of destination involvement on tourists' revisit intention and defined it as a dual dimension construction (psychological and behavioral involvement) and measured each dimension by three elements: attraction; performance risk; and financial risk, but it seems that behavior involvement concerns dependent variables and destination outcome that tourists involved, such as on-site tourism activities, pre-travel planning, spending, and so forth (Hu, 2003). Relevant literature about destination involvement provides the basis and inspiration for this study. Involvement theory is shown to be an important concept in the understanding of consumers' behavior and psychology. At present, the concept has been empirically explored in limited contexts, such as rural tourism, self-driving tourism, ecotourism, and museum tourism; little has been found in wildlife tourism settings. Furthermore, research on the influences and mechanism of tourists' involvement on travel behavior are relatively insufficient (Huang et al., 2014). The research on tourists' destination involvement is meaningful to understand wildlife tourists' satisfaction, behavior, and performance within sites and revisit willingness and it is conducive to the management of destination and the formulation of marketing strategy.

The activities in this paper only occur within the sites of wildlife tourism. Site is one of the parts of destination. It is relatively raised with "space" (Tuan, 1997). Without a cultural component, space is precisely expressed by vector in which there are only geographic location and physical form; however, site has significant cultural factors. Therefore, based on the literature review, this study defines place involvement as a special relationship between tourists and place. There are two key meanings: first, place involvement refers to tourists' psychological and behavioral activities involved in certain wildlife tourism places; and second, it refers to the wildlife tourists' involvement in the wildlife places.

6.3 METHODOLOGY

6.3.1 CASE STUDY: THE DOLPHIN DISCOVERY CENTRE, BUNBURY, AUSTRALIA

The Dolphin Discovery Centre (DDC) is located in downtown Bunbury, ~180 km south of Perth, with a population of 31,000. Bunbury is the second largest city in Western Australia and provides an opportunity for watching dolphins in their wild habitat (City of Bunbury, 2015). As a non-profit organization, the DDC opened in 1994 and fulfills its mission of protecting dolphins and their habitats by carrying out tourism, education, and scientific research. Its annual visitor flow is about 60,000—40,000 oversea visitors and 20,000 Australian visitors (Newsome et al., 2005). There is a number of wildlife tourism activities offered at the DDC: dolphin watching at an interactive area of beach; dolphin watching on a pelagic pleasure-boat; a swimming with dolphins tour and the opportunity to volunteer.

The DDC was selected as a case study for three key reasons: first of all, non-resource-consumptive wildlife tourism has a long history in Australia and develops very well, which not only meets visitors' need of close contact with wildlife in the wild habitats, but also it is good for the protection of wildlife and their habitats. Its developmental experience is worthy of our learning and reference. Second, as one of the brands of Australian wildlife tourism, dolphins are famous and are widely distributed in the oceans. Finally, the DDC has a wildlife product, which is attractive to a large number of visitors from all over the world; it is well known and may be regarded as representative of other wildlife settings.

6.3.2 THE DEVELOPMENT OF THE PLACE INVOLVEMENT SCALE

The main data collection method used in this research is questionnaire and the data analysis methods are descriptive statistical analysis and ANOVA analysis. In terms of the measurement of place involvement, from the perspective of product factor, site involvement is the major measure standard of tourists within wildlife tourism. The draft of scale items is based on literature review, including Zaichkowsky (1994), Laurent and Kapferer (1985), and Lehto et al. (2004), with total 12 items. These items are highlighted below:

1. Dolphin travel is valuable;
2. I like dolphin tourism very much;
3. I like watching dolphins in the wild habitat for a long time and within a close range;
4. I like destination tourism with travel guide,
5. Someone says that mother tongue is very important in dolphin travel;
6. Some values obtained from the spending of dolphin travel are very important to me;
7. The cost of dolphin travel cannot be too high, which is important to me,
8. How long it will take to start planning the trip before departure?
9. I am generally satisfied with the dolphin travel,

10. Participating in dolphin travel is worthwhile;
11. I will probably revisit this place to see the dolphins within the next five years;
12. I am willing to revisit this place to see the dolphins within the next five years.

Given that, importance, symbolic meaning, and entertainment are the three key dimensions of measuring place involvement for wildlife tourism sites, four experts from universities in America, Australia, and China in wildlife tourism were invited to evaluate the content of these items. In the process, experts were required to evaluate whether the items could measure the target dimension effectively. The item was deleted once two experts considered it improper. After consulting experts' opinions, items 1, 2, 3, 9, and 12 were left.

And then, six postgraduate students majoring human geography from the Center of Recreation of Tourism Research in Peking University (China) and 10 tourists who had experience of wildlife tourism were invited to attempt the questionnaire in order to further modify some expressions of these measurement items. Finally, five items were left, namely *dolphin travel is valuable; I like dolphin tourism very much; I like watching dolphins in the wild habitat for a long time and within a close range; I am generally satisfied with the dolphin travel*; and *I will probably revisit this place to see the dolphins within the next five years*. Through in-depth interview, conceptualized operation, and literature review, available items in various dimensions of the scale are selected and appropriate structure variables of the items are basically decisive to the validity of content measured (Ping, 1993). Therefore, the scale in this paper enjoys relatively good content validity. Each item utilizes a five-point Likert scale ranging from strong agreement, agreement, neutral opinion, disagreement, and strong disagreement.

Cronbach's α, the most commonly used method of examining reliability, was applied to this scale with a result of 0.636 which is higher that the threshold value of 0.6 and therefore passes this test of reliability.

6.3.3 QUESTIONNAIRE DISTRIBUTION

This research was greatly supported and helped by the research team of Murdoch University in Australia and DDC. The questionnaire underwent a Human Ethics application through Murdoch University and received approval. The empirical works were completed over January 4–13, 2014 through three different approaches. First, in order to potentially capture all visitors attending the DDC, a questionnaire reception area was set in the rest zone of the center; all tourists were welcomed to complete the questionnaire in a volunteer capacity. Second, several investigators randomly selected visitors who had interacted with dolphins along the beach; these surveys were completed and collected on site between the peak time of 08:00–11:00 in the morning and 13:00–16:00 in the afternoon. The final approach involved staff and volunteers targeting tourists who had been on a cruise or a swim tour. A small gift (panda souvenir) was used for appreciation and encouragement in regard to participation (response rate 90%) in the survey. Finally, 348 questionnaires were collected according to the strategy described above. After removing questionnaires with incomplete information, 304 valid questionnaires were collected and were used for analysis.

6.3.4 DATA ANALYSIS

Data from the 304 questionnaires were input into Excel and SPSS 17.0. The main statistical methods of analysis were K-means cluster, independent sample t-test, ANOVA, and Scheffe's post hoc test.

6.4 RESULTS AND ANALYSIS

6.4.1 THE CHARACTERISTICS OF SAMPLE DEMOGRAPHY

The sample of respondents reflected the following characteristics. The visitors were predominately Australian (217persons, 71%) with the United Kingdom (27 persons), France (13 persons), Germany (8 persons), Switzerland (8 persons), and China (7 persons including 2 Taiwanese) followed. In total, there were 26 countries from where visitors originated. The sample was predominately female (63%) with 65% of visitors aged between 20 and 50 years and 62% having an education level of a bachelor degree or higher. In terms of travel parties, 70% were accompanied by other family members with 43% travelling as couples with two or more children.

In terms of travel behavior, 68% had their travel carefully arranged to watch wildlife, 47% used the network as information source, and 38%

noted that they had watched dolphins six times or more previously. Their aims for dolphin interaction included 27% who simply wished to view dolphins and 20% who were seeking a deeper experience by swimming with dolphins. Expenditure at the site was noted as under $AUD100 by 61% of respondents (Table 6.1).

6.4.2 OVERVIEW OF BEHAVIOR CHARACTERISTIC OF PLACE INVOLVEMENT

TABLE 6.1 Characteristics of DDC Wildlife Tourist's Behavior.

Variable	Item	Frequency	Percent (%)
Viewing times	1 time	46	14.5
	2–3 times	85	28.0
	4–5 times	52	17.0
	More than six times	116	38.0
Travel companion	Single person	22	7.2
	Family members	213	70.1
	Friends	63	20.7
	Team organization	6	2.0
	Guides	0	0.0
	Business partners	1	0.3
	Others	3	1.0
Duration of stay	Less than one night	103	33.9
	1 night	63	20.7
	2 nights	37	12.2
	3 nights	25	8.2
	4–5 nights	15	4.9
	More than six nights	54	17.8
Destination consumption	Under $AUD100	35	61.0
	$AUD100–199	58	19.0
	$AUD200–499	36	12.0
	$AUD 500–999	12	4.0
	Above $AUD1000	12	4.0

TABLE 6.1 *(Continued)*

Variable	Item	Frequency	Percent (%)
Expect next visit	Watch dolphins	73	24.0
	Photo shoot closely	76	25.0
	Observe them in a short distance	155	51.0
	Feed dolphins and take pictures with them	112	37.0
	Take part in voluntary projects	55	18.0
	Take part in projects of donating money and adopting dolphins	61	20.0
	Take part in the project of guiding dolphins	30	10.0
	Other forms	0	0.0
Participate in activities	Watch animals from a long distance	268	88.0
	Photo shoot closely	234	77.0
	Observe them closely/swim with dolphins	61	20.0
	Feed dolphins and take pictures with them	6	2.0
	Take part in voluntary projects	15	5.0
	Take part in projects of donating money and adopting dolphins	67	22.0
	Take part in the project of guiding dolphins	30	10.0
	Take part in associations or organizations of wildlife conservation	5	1.6
	Other activities	0	0.0
Behavior disposition after traveling	Tell relatives and friends	232	76.3
	Invest more money for wildlife conservation and tourism	53	17.4
	Share travel experiences online	83	27.3
	Take part in associations of wildlife conservation	5	1.6
	Take part in volunteer tourism	149	49.0
	Other	13	4.3

Note: Frequencies reflect the number of responses to the survey question. Percentages reflect the number of the total sample providing the response.

The majority of respondents from the DDC sample are repeat visitors. As indicated in Table 6.1, only 15% of those surveyed indicated that this was the first time they had viewed dolphins in the wild. As for the duration of stay, 34% of the sample did not stay overnight in Bunbury, 21 responses stayed overnight, 25% stayed for between 2 and 5 nights, and 18% stayed six nights or longer.

Of the 153 visitors (*n*) who provided a response to the survey question relating to expenditure, 23% spent less than $AUD100 and 61% spent less than $AUD200, whilst 8% spent more than $AUD1000. The key areas of expenditure included visitors who chose the cruise ($35AUD per person), the swim with dolphins tour ($AUD165 per person), or donated to the DDC (22%).

The survey question eliciting responses to post behavior indicated that 76% of respondents would share their experience with friends and relatives, 27% of visitors said that they would share their experiences online and 49% indicated a willingness to take part in volunteer tourism. Respondents' expectations of future wild dolphin tourism activities indicated that 51% would like a deeper experience of watching dolphins closely and 37% indicated that they would like to experience feeding dolphins.

6.4.3 OVERALL EXTENT OF PLACE INVOLVEMENT SCALE

According to respondents' scores of involved factors within different places and Q clustering analysis on all tourists, through K-Means clustering method and repeated tests, three clusters were developed with mean scores of 2.33, 3.58, and 4.4 (Table 6.2). By one-way variance analysis, these three categories have significant differences on five items, which indicates the clustering effect is good. Based on the results of cluster analysis, this paper proposes the classification criteria of place involvement as follows:

- Where the mean value of place involvement scale is less than or equal to 2.6, the respondent is placed in the *shallow place involvement (SPI)* category;
- Where the mean value of place involvement scale is between 2.6 and 4; the respondent is placed in the *medium place involvement (MPI)* category;

TABLE 6.2 Analysis of Demographic Characteristics of Three Types of Place Involvement.

Measurement Dimension	Item	DPI ($n = 201$)		MPI ($n = 97$)		SPI ($n = 6$)	
		Frequency	Percent (%)	Frequency	Percent (%)	Frequency	Percent (%)
Gender	Male	67	33.33	42	43.30	2	33.33
	Female	134	66.67	55	56.70	4	66.67
Age	Less than 16	13	6.47	5	5.15	0	0.00
	16–19	11	5.47	6	6.19	0	0.00
	20–29	37	18.41	30	30.93	3	50.00
	30–39	41	20.40	26	26.80	3	50.00
	40–49	41	20.40	18	18.56	0	0.00
	50–59	35	17.41	8	8.25	0	0.00
	More than 60	23	11.44	4	4.12	0	0.00
Education background	Under Junior school	1	0.50	1	1.03	0	0.00
	High school	72	35.82	35	36.08	3	50.00
	Vocational school	32	15.92	15	15.46	0	0.00
	Bachelor	65	32.34	28	28.87	2	33.33
	Master	17	8.46	13	13.40	0	0.00
	Doctor	10	4.98	3	3.09	1	16.67

TABLE 6.2 *(Continued)*

Measurement Dimension	Item	DPI ($n = 201$)		MPI ($n = 97$)		SPI ($n = 6$)	
		Frequency	Percent (%)	Frequency	Percent (%)	Frequency	Percent (%)
Income (Y/$)	Less than 10,000	26	12.94	17	17.53	0	0.00
	10,000–19,000	11	5.47	9	9.28	1	16.67
	20,000–49,999	23	11.44	14	14.43	0	0.00
	50,000–99,000	56	27.86	23	23.71	1	16.67
	100,000–199,999	60	29.85	21	21.65	2	33.33
	More than 2,00,000	4	1.99	4	4.12	0	0.00
	Blank	21	10.45	9	9.28	2	33.33
Family background	Single	42	20.90	27	27.84	0	0.00
	Couple with no child	43	21.39	24	24.74	3	50.00
	Couple with one child	16	7.96	12	12.37	1	16.67
	Couple with two or more children	96	47.76	34	35.05	2	33.33

- Where the mean value of place involvement scale is equal to or greater than 4, the respondent is placed in the *deep place involvement* (DPI) category.

The mean value of the DDC sample is 4.1, suggesting that this wildlife tourism site belongs in the DPI category. Results from the survey show the number of the DPI respondents is 201 (61%), number of the MPI is 97 (32%) with number of the SPI respondents is only 6 (2%).

The demographic characteristics of MPI include that the male proportion is higher than the female's; respondents among 20–29 take up higher proportion; 60% of respondents have received undergraduate degree and above; over 74% annual income are less than 50,000 $AUD; and over 66% are single or childless. Therefore, the majority of MPI is those single young people with less income and medium education background, or even a little bit higher.

6.4.4 DEMOGRAPHIC DETERMINATES OF PLACE INVOLVEMENT

One-way ANOVA has been used here to analyze the differences between the sociological characteristics of wildlife tourists' place involvement. Scheffe's post hoc has been applied to verify the results achieved by one-way ANOVA results. Among them, the variance analysis of two variables in sociological characteristics is examined by an independent sample t-test. The main conclusions of the test are as follow:

By independent sample t-test, it is concluded that different genders do not have significant differences ($p > 0.05$) on different involvement variables, while nationality have significant differences ($p < 0.05$) on the scale item of revisit in the next five years in which the number of Australian tourists is significantly higher than that of non-Australian tourists.

Different age groups in four variable items have significant impacts on the difference of place involvement ($p < 0.05$). Two factor dimensions had significant differences with Scheffe's post hoc test afterwards, as shown in Table 6.3. The mean value of the subgroup with age amongst 20–29 has relatively low scores, while the group aged amongst 50–59 had the highest scores (Table 6.3).

TABLE 6.3 Variance Analysis of Age Group on Place Involvement.

Items Age	Value	Different age group (N = 304)							F	p	Scheffe's post hoc test
		<16 N = 18	16–19 N = 16	20–29 N = 69	30–39 N = 69	40–49 N = 59	50–59[6] N = 43	≥60 N = 26			
Q1	Mean	3.889	4.000	3.700	3.786	4.102	4.186	4.385	2.94	0.008	
	SD	0.832	0.791	0.938	1.062	0.995	0.906	0.571			
Q2	Mean	4.222	4.059	3.986	3.986	4.220	4.429	4.407	4.01	0.001	
	SD	0.732	0.659	0.752	0.940	0.744	0.668	0.501			
Q3	Mean	4.389	4.235	4.200	4.157	4.085	4.047	4.231	1.23	0.289	
	SD	1.037	0.664	0.972	0.828	0.915	1.045	0.710			
Q4	Mean	4.500	4.353	4.086	4.257	4.424	4.628	4.482	4.38	0.000	3<6
	SD	0.618	0.493	0.737	0.755	0.675	0.536	0.509			
Q5	Mean	3.889	4.000	3.386	3.886	4.119	4.302	4.074	2.80	0.010	3<6
	SD	1.132	0.866	1.107	0.910	1.035	0.914	0.997			
Sample	Mean	4.178	4.129	3.872	4.014	4.190	4.318	4.316			
	SD	0.870	0.695	0.901	0.899	0.873	0.814	0.658			

Note: *Q1: Dolphin travel is valuable; Q2: I like dolphin tourism very much; Q3: I like watching dolphins in the wild habitat for a long time and within a close range; Q4: I am generally satisfied with the dolphin travel; Q5: and I will probably revisit this place to see the dolphins within the next five years.*

Education backgrounds have significant difference ($p < 0.05$) in satisfaction variable (Q4) while by Scheffe's post hoc analysis, the education background groups have not. Among them, the highest mean value is doctoral degree group C6 = 4.26 and the lowest is the senior high school with mean value of 4.01. Through examination, all the scale mean values of place involvement are over four, belonging to DPI.

Different income levels have no significant differences ($p > 0.05$) on all items. Through examination on all the scale mean values of place involvement, it is found that respondents earning between $AUD10,000–20,000 and $AUD50,000–100,000 are over four with those earning $AUD100,000–200,000 having the strongest level of place involvement with a mean value of 4.2. All three groups reflect a DPI. However, it is interesting to note that those respondents with earnings of $AUD200,000 or more reflected the lowest level of place involvement with mean value of 3.8.

Different family backgrounds have significant differences ($p < 0.05$) on the item of revisit (Q5) while there is no significant difference on other items as shown in Table 6.4. Through examination of all scale item mean values of place involvement, it was concluded that couples with one child or more than one, and single respondents have the highest scores with mean value of 4.2, 4.0, and 4.0, respectively. The mean values of these three groups' are bigger than four, so they are DPI. The place involvement of those couples without a child is just less than four, mean value is 3.994 which is classifies to the MPI (Table 6.4).

TABLE 6.4 Variance Analysis on Varied Family Background in Place Involvement.

Items	Value	Different family background ($N = 650$)				F	p	Post test
		Single $n = 67$	C1 $n = 70$	C2 $n = 29$	C3 $n = 131$			
Q1	Mean	3.812	3.771	4.069	4.083	2.332	0.074	
	SD	0.989	0.981	0.923	0.917			
Q2	Mean	4.188	4.071	4.138	4.168	0.314	0.815	
	SD	0.692	0.804	0.789	0.815			
Q3	Mean	4.246	4.243	3.828	4.136	1.746	0.158	
	SD	0.864	0.955	1.104	0.845			

TABLE 6.4 *(Continued)*

Items	Value	Different family background ($N = 650$)				F	p	Post test
		Single $n = 67$	C1 $n = 70$	C2 $n = 29$	C3 $n = 131$			
Q4	Mean	4.304	4.214	4.517	4.394	1.763	0.154	
	SD	0.734	0.720	0.509	0.674			
Q5	Mean	3.754	3.671	3.690	4.129	4.194	0.006	2<4
	SD	0.976	0.989	1.466	0.960			
Sample	Mean	4.061	3.994	4.048	4.182			
	SD	0.851	0.890	0.958	0.842			

Note: *Q1: Dolphin travel is valuable; Q2: I like dolphin tourism very much; Q3: I like watching dolphins in the wild habitat for a long time and within a close range; Q4: I am generally satisfied with the dolphin travel; Q5: and I will probably revisit this place to see the dolphins within the next five years. C1: Couple, no child; C2: Couple, one child; C3: Couple, two children or more.*

6.4.5 TOURIST BEHAVIOR SIGNIFICANTLY AFFECTS PLACE INVOLVEMENT

Travel behaviors and satisfaction have significant differences in the place involvement with independent sample t-tests. Travel behaviors include length of stay, source of information, and travel companion. Regarding length of stay, it had significant differences in place involvement generally ($p < 0.05$); the respondents staying less than one night are significantly lower than respondents staying over two nights, as shown in Table 6.5.

Place consumption had significant differences in the item of being very satisfied with wildlife tourism generally ($p < 0.05$).The number of groups spending less than $AUD100 is significantly lower than that spending more than $AUD100.

Source of information had significant differences in the item that I like watching the wild animals in the wild ($p < 0.05$); place involvement mainly comes from the personal travel experience, which is significantly lower than other groups.

Travel companion had significant differences in the item wildlife tourism is valuable ($p < 0.\ 05$), place involvement of traveling alone is significantly lower than those having companion.

Satisfaction had significant differences in the item that I especially like wildlife tourism ($p < 0.05$); place involvement of these groups with high satisfaction is significantly higher than these groups with dissatisfaction.

TABLE 6.5 Variance Analysis of Tourism Behavior.

Scale item	Item	Groups	Frequency	Mean value	SD	*F*	Sig. diff.
Very satisfied with wild-life tourism generally	Duration of stay	Less than one night	103	4.233	0.819	4.401	0.037**
		Over two nights	201	4.298	0.600		
	Place consumption	Less than 100	179	4.257	0.758	4.930	0.027**
		More than 100	125	4.464	0.547		
I like watching wildlife in the wild	Source of information	Personal experience	35	4.028	1.071	3.959	0.048**
		Others	269	4.179	0.881		
Wildlife tourism is valuable	Travel companion	Travel alone	22	3.682	1.323	8.598	0.004**
		Accompanied	281	3.975	0.919		
I especially like wildlife tourism	Satisfaction	Dissatisfaction	26	3.81	0.958	5.346	0.021**
		Satisfaction	278	4.18	0.900		

Note: **Represents variables at the level of <0.05 are significant.

6.5 CONCLUSIONS AND DISCUSSION

Place involvement is a vital issue in wildlife tourism. Understanding tourists' place involvement and differences is conducive to the analysis of tourists' decision-making, destination behavioral characteristics, and destination management. There are many wild animals in Australia those tourists enjoy interacting with. Dolphin watching is a flagship activity in Australian terms of wildlife tourism products. This paper takes the Dolphin Discovery Centre, Bunbury, Western Australia as a case study and carries out an analysis of place involvement illustrated in the wildlife tourism context. A variety of empirical quantitative analysis methods, including the construction of a measurement scale for place involvement, K-means cluster, independent sample t-test, ANOVA, and Scheffe's post hoc test were used. The main conclusions are summarized below.

First, place involvement is defined and the theory is examined in context. Place involvement is a unique relationship between tourist and place, which includes two key perspectives. First, it refers to tourists' psychological and behavioral activities involved in certain wildlife tourism places and second, with place as the involvement object, place refers to the wildlife tourists' involvement in the places.

Place involvement has behavioral and psychological dimensions. In terms of this case study, DDC respondents have several ways of interacting with dolphins in their wild habitat. These opportunities include, watching dolphins at the interactive beach area; watching dolphins on cruises; swimming with dolphins in the ocean; and being volunteers. Respondents' behavioral characteristics include most tourists having watched dolphins more than once. Their travel companions are family members and the whole process has relatively low place consumption in terms of expenditure. The proportions of no destination stay and staying more than six nights are relatively high. Many tourists are willing to watch dolphins from a long distance or observe them closely. In the post-travel behavioral tendency, most respondents are willing to share experience with friends and relatives as well as to take part in volunteer tourism. In terms of the expectation of next activities, the desire to observe dolphins closely and feeding them accounts for a high proportion.

Second, through K-means clustering analysis, place involvement was classified into three categories including SPI, medium place involvement, and DPI. According to partition criterion, when the mean value of place involvement scale is less than or equal to 2.6, it belongs to SPI; when the mean value of place involvement scale is between 2.6 and 4, it belongs to MPI; when the mean value of site involvement place is larger than or equal to 4, it belongs to DPI. The overall sample in this case study reflected a mean score of 4.1, suggesting that the DDC as a wildlife tourist site attracts visitors that are seeking to DPI.

Through the application of independent sample t-tests, ANOVA, and Scheffe's post hoc approach, this paper examined the differences in demographic characteristics and geographic characteristics associated with place involvement. It was found that nationality, age, and family background had significant difference on place involvement ($p < 0.05$), while gender, educational background did not ($p < 0.05$). The majority of DPI in wildlife tourism is female with relatively high income and educational background. Single people and couples with more than one child also take up a high proportion in the DPI. In MPI, the majority appear to be those who are young with less income and a general educational level.

By independent sample t-tests, this paper also examines the differences of various travel behaviors and satisfaction of groups in relation to place involvement. It is suggested that duration of stay, place consumption, sources of information, travel companions, and satisfaction have significant differences ($p < 0.05$).

In the process of wildlife tourism, tourists' management is very important where some desired tourist activities and behaviors may lead to negative effects. These include some behaviors such as having close contact with animals or feeding them, which could result in the spread of disease or threat the personal security (Newsome et al., 2005). Although tourists hope to have close contact with wild animals, such contact may disturb or terminate animals' normal behavior or push them give up some of their living conditions (Ma & Cheng, 2008). In the case of the DDC, some management measures on tourists' place involvement are implemented. There is the establishment of a clear area designated for beachside dolphin interaction. Volunteers are encouraged to provide clear guidelines and interpretive information about the dolphin wildlife with the aim of providing useful dolphin-related knowledge so that visitors may have better understanding of their own behavior on the dolphins in an attempt to reduce interference to these wild dolphins. Furthermore, volunteers serve an important role in wildlife tourism. The DDC has cooperated with Murdoch University and other research institutions that conduct scientific research on dolphins in terms of their physical habits, reproduction, and behavioral characteristics, providing important technical support for the development of wildlife tourism activities.

Environmental education is an important part of DDC tourism management. The DDC has a focus on dolphin research and tourism interaction, which provides both a revenue and political support opportunity for their research endeavors. Research in dolphin conservation underpins their tourism activities where long-term observation and scientific research contribute to their choices of activities. Limited beach feeding has been examined in terms of the health and habituation of dolphins; dolphin viewing cruise tours have strict requirements in terms of dolphin interaction with the boats, and swimming with dolphins (Lee & O'Neill, 2000). It is necessary to take measures such as the establishment of visitor center, the issue of brochures, various means of interpretation, and education. Meanwhile, grasping some basic information, including the requirement of wildlife tourism activities, environmental awareness, and tourists' behaviors, the number of visitors, time, and spatial distribution is also important (Ma & Cheng, 2008).

The limitations of this research and hints for future research are as following: first, this research only examined one species in one setting; the factors affecting involvement can be extended in regard to the examination of more species in more settings thus helping to verify the above findings. Such work could strengthen research on tourists' place involvement and influential factors of more species and habitats. Second, according to the literature, place involvement may be influenced by tourists' personal traits (Gursoy & Gavcar, 2003; Huang et al., 2014) and is related to tourists' risk perception and preference scale (Huang et al., 2014). Risk perception or environmental attitude could be examined with the relationship with involvement in the future. In addition, China has rich wildlife tourism resources. Wild animals such as the giant panda, the golden monkey, and the Asian elephant have strong value of appreciation and great market potential and these species may gain a greater focus for wildlife tourism, with specific attention to the domestic Chinese market. Future research could be targeted with Chinese species. At last but not the least, this paper suggests that long-term monitoring mechanisms in wildlife tourism should be established and some multi-disciplinary social science research on tourists' psychology, attitude, and behavior in several key regions should be carried out in order to lay a solid foundation to the sustainable development of wildlife tourism. This DDC case study provides a wealth of information for the exploration of future wildlife tourism planners.

KEYWORDS

- **wildlife tourism**
- **non-consumptive**
- **place involvement**
- **Australia**
- **dolphin**

REFERENCES

Beaton, A. A.; Funk, D. C. An Evaluation of Theoretical Frameworks for Studying Physically Active Leisure. *Leis. Sci.* **2008,** *30* (1), 53–70.

Boyle, S. A.; Samson, F. B. Effects of Nonconsumptive Recreation on Wildlife: A Review. *Wild. Soc. Bull.* **1985,** *13* (2), 110–116.

Cai, L. A.; Feng, R. M.; Breiter, D. Tourist Purchase Decision Involvement and Information Preferences. *J. Vacat. Mark.* **2004,** *10* (2), 138–148.

Carneiro, M. J.; Crompton, J. L. The Influence of Involvement, Familiarity, and Constraints on the Search for Information about Destinations. *J. Travel Res.* **2010,** *49* (4), 451–470.

Catlin, J.; Jones, R. Whale Shark Tourism at Ningaloo Marine Park: A Longitudinal Study of Wildlife Tourism. *Tourism Manage.* **2010,** *31* (3), 386–394.

City of Bunbury. http://www.bunbury.wa.gov.au (accessed Sep, 2015).

Cong, L.; Newsome, D.; Wu, B.; Morrison, A. M. Wildlife Tourism in China: A Review of the Chinese Research Literature. *Curr. Issues Tourism.* **2014,** *8,* 1–23.

Cong, L.; Wu, B.; Morrison, A. M.; Wang. M.; Shu, H. Analysis of Wildlife Tourism Experiences with Endangered Species: An Exploratory Study of Encounters with Giant Pandas in Chengdu, China. *Tourism Manage.* **2014,** *40* (2), 300–310.

Cong, L.; Wu, B.; Li, J. A Literature Review on Overseas Wildlife Tourism Research. *Tour. Trib.* **2012,** *5* (23), 57–65.

Curtin, S. C. Managing the Wildlife Tourism Experience: The Importance of Tour Leaders. *IJTR.* **2010,** *12* (3), 219–236.

Duffus, D. A.; Dearden, P. Non-Consumptive Wildlife-Oriented Recreation: A Conceptual Framework. *Biol. Cons.* **1990,** *53* (3), 213–231.

Gursoy, D.; Gavcar, E. International Leisure Tourists' Involvement Profile. *Ann. Tour. Res.* **2003,** *30,* 906–926.

Havitz, M. E.; Dimanche, F. Leisure Involvement Revisited: Conceptual Conundrums and Measurement Advances. *J. Leis. Res.* **1997,** *29* (3), 245–278.

Havitz, M.; Dimanche, F. Leisure Involvement Revisited: Drive Properties and Paradoxes. *J. Leis. Res.* **1999,** *31,* 122–149.

Houston, M. J.; Rothschild, M. L. Conceptual and Methodological Perspective in Involvement. In *Research Frontiers in Marketing: Dialogues and Directions;* American Marketing Association: Chicago, 1978; pp 184–187.

Hu, B.; Yu, H. Segmentation by Craft Selection Criteria and Shopping Involvement. *Tour. Manag.* **2007,** *28,* 1079–1092.

Hu, B. The Impact of Destination Involvement on Travelers' Revisit Intentions (Unpublished Doctoral Dissertation). Purdue University: West Lafayette, IN, 2003.

Huang, L.; Gursoy, D.; Xu, H. Impact of Personality Traits and Involvement on Prior Knowledge. *Ann. Tour. Res.* **2014,** *48,* 42–57.

Hwang, S. N.; Lee, C.; Chen, H. J. The Relationship Among Tourists' Involvement, Place Attachment and Interpretation Satisfaction in Taiwan's National Parks. *Tour. Manage.* **2005,** *26* (2), 143–156.

Huang, Z.; Yuan, L.; Yu, Z. Forecasts of Tourist Flow Features in Ecotourism Area: A Case Study of Yancheng David's Deer Ecotourism Area. *Acta Geogr. Sin.* **2007,** *62* (2), 1277–1286. (In Chinese).

Huang, Z.; Yuan, L.; Yu, Z.; Wu, J.; Zhou, N. The Spatio-Temporal Evolution and Characteristics Analysis of Tourist Flow in Ecotourism Area: A Case Study of Yancheng Ecotourism Area for David's Deer. *Geogr. Res.* **2008,** *27* (1), 55–64. (In Chinese).

Kim, S.; Scott. D.; Crompton, J. L. An Exploration of the Relationships among Social Psychological Involvement, Behavioral Involvement, Commitment, and Future Intentions in the Context of Birdwatching. *J. Leis. Res.* **1997,** *29* (3), 320–329.

Laurent, G.; Kapferer, J. Measuring Consumer Involvement Profiles. *J. Market. Res.* **1985,** *22* (2), 41–53.

Lee, D.; O'Neill, F. Best Practice Dolphin-Tourism Interaction Management. *Touristics.* **2000,** *16* (1), 15–18.

Lehto, X. Y.; O'Leary, J. T.; Morrison, A. M. The Effect of Prior Experience on Vacation Behavior. *Ann. Tour. Res.* **2004,** *31* (4), 801–818.

Li, L. Research of Birdwatching Tourists Behavior Take the Birdwatching Tourists in Beijing For Example (Master's Thesis). Beijing Forestry University, Beijing (In Chinese), 2009.

Ma, J.; Cheng, K. Impacts of Ecotourism on Wildlife in Nature Reserves: Monitoring and Management. *Acta Ecologica Sinica.* **2008,** *28* (6), 2818–2827. (In Chinese).

McGehee, N. G.; Yoon, Y.; Cárdenas, D. Involvement and Travel for Recreational Runners in North Carolina. *J Sport Manage.* **2003,** *17,* 305–324.

Newsome, D.; Dowling, R.; Moore, S. *Wildlife Tourism;* Channel View Publications: Britain, 2005; 16–22, 209–212.

Ping, R. A. The Effects of Satisfaction and Structural Constraints on Retailer Exiting, Voice, Loyalty, Opportunism and Neglect. *J. Retail.* **1993,** *69* (3), 320–352.

Prayag, G.; Ryan, C. Antecedents of Tourists' Loyalty to Mauritius the Role and Influence of Destination Image, Place Attachment, Personal Involvement, and Satisfaction. *J. Travel Res.* **2012,** *51* (3), 342–356.

Sindiyo, D. M.; Pertet, F. N. Tourism and its Impact on Wildlife Conservation in Kenya. *Indus. Environ.* **1984,** *7* (1), 14–1911.

Tuan, Y. F. *Space and Place: The Perspective of Experience.* University of Minnesota Press: Minneapolis, 1977; pp 8–18, 118–1481.

Wang, H. *Impacts and Evaluation of Wildlife Recreation-a Case Study of Birdwatching Ecotourism* (Doctoral Dissertation). Beijing Forestry University: Beijing (In Chinese), 2008.

Xu, H. Development and Problems of Non-Consumptive Wildlife Tourism in China. *Geogr. Geoinf. Sci.* **2004,** 20, 82–86, 90.

You, X.; Dai, N. Development and Countermeasure of Poyang Lake Birdwatching Ecological Tourist. *Jiangxi Sci.* **2010,** *28* (6), 867–870. (In Chinese).

Yuan, L.; Yu, Z.; Huang, Z.; Gu, Q. The Functional Mechanism of Tourism-generating Destinations' Social-Economic Factors to Tourist Numbers in Eco-Tourism Scenic Spots - A Case Study of Yancheng Eco-Tourism Area for David's Deer. *Hum. Geogr.* **2007,** *22* (6), 120–123. (In Chinese).

Yuan, L.; Yu, Z.; Huang, Z.; Gu, Q. An Analysis of Multi-Fluctuation Characteristics and Functional Process of Tourist Flow: A Case Study of Yancheng David's Deer Ecotourism Area. *Tourism Trib.* **2009,** *24* (7), 27–32. (In Chinese).

Zaichkowsky, J. Measuring the Involvement Concept. *J. Cons. Res.* **1985,** *12,* 341–352.

Zaichkowsky, J. L. The Personal Involvement Inventory: Reduction, Revision, and Application to Advertising. *J. Advertising.* **1994,** *23* (4), 59–70.

Zhang, H.; Lu, L. Impacts of Tourist Involvement on Destination Image: Comparison between Inbound and Domestic Tourists. *Acta Geogr. Sin.* **2010,** *5* (12), 1613–1623. (In Chinese).

CHAPTER 7

SOCIAL PREFERENCE OF LOCAL COMMUNITIES TOWARDS WILDLIFE TOURISM IN THE INDIAN HIMALAYAS

PARIVA DOBRIYAL, RUCHI BADOLA, and SYED AINUL HUSSAIN*

Wildlife Institute of India, Dehradun, Uttarakhand, India

**Corresponding author. E-mail: hussain@wii.gov.in*

CONTENTS

ABSTRACT

This chapter presents the attitude of mountain communities toward wildlife tourism as a livelihood option in the Nanda Devi Biosphere, India. The Reserve is a world heritage site with significant scenic, cultural, and religious value. A total of 764 interviews were conducted from 22 selected villages using semi-structured questionnaire in randomly selected households (HH). Respondents were asked to assign scores on the scale of 4–1 (four being the highest and one the lowest) to wildlife tourism in the area. About 36% HH had positive attitude toward wildlife tourism in the area and assigned scores to it. Only few respondents supported wildlife tourism as it is related to conservation of wildlife leading to increased incidence of negative interaction between local communities and wildlife. Average income was higher for the HH involved in tourism than the HH involved in other livelihood activities. It was observed that younger members of the community were more positive toward wildlife tourism related activities in the area. Similarly, HH with higher average year of schooling and without a regular income supported wildlife tourism in the area as they see it as a livelihood opportunity. Assessment of the attitude and behavior of local communities toward sustainable nature based and wildlife tourism can be used to understand the opportunities and threats that the conservation and management programs are going to face during implementation in a particular area. These issues can be dealt by addressing the issues highlighted by the local communities and considering the value local people put on flow of environmental services.

> **"Wildlife tourism....proven its importance in enhancing the livelihoods of many local communities"**

7.1 INTRODUCTION

Tourism industry has become the major economic sector for most developing countries and societies (Eraqi, 2007). Tourism enhances the purchasing power of the economically weak by providing better livelihoods (Látková & Vogt, 2012). Tourism has diversified the economic status of the underdeveloped societies and has encouraged them to adopt new economic development strategies on the basis of natural and cultural resources available to them (Garrod, 2006). The recent tourism development plans for nature-based tourism destinations have raised concerns about how sustainable land use targets can be achieved. Researchers such as Eraqi (2007) and Ritchie (1993) have emphasized that residents of a tourist destination should not be taken for granted and must be responsive to the planning and implementation efforts of public–private partnerships.

Given the mutually supportive relationship between nature and human beings, wildlife tourism depicts the strong interrelationship between the two. The attitude toward wildlife tourism has the potential to positively impact tourism in terms of appreciation of wildlife and environment as a whole and aiding its conservation (Ballantyne et al., 2011). Nature tourism unravels the anthropomorphic connections to wildlife and encourages a deep sense of well-being and spiritual fulfillment and psychological benefits (Curtin, 2009). Wildlife tourism not only plays a significant role in promoting recreational, aesthetic, and cultural values but has also proven its importance in enhancing the livelihoods of many local communities in terms of benefits from biomass production or park funds diverted to local villages by state agencies and revenue from wildlife tourism (Sekhar, 2003). Conservationists look forward to and encourage the active participation from local communities in management of nature tourism. At this juncture where population is growing at a rapid pace and so are the demands, it is crucial to encourage and actualize the concept of ecotourism or sustainable tourism for habitat and wildlife conservation so that the future generations do not miss out on what we have taken for granted.

In India, high densities of people live around protected areas (PAs) and share the natural resources with wild animals (Karanth & Nepal, 2012). Benefits from resources provided by the natural ecosystem to local communities are generally substantial and decrease on the declaration of the area as a PA (Sandbrook, 2010; de Pinho et al., 2014). This poses a challenge for the conservation goals and needs of local residents, and is the major task for the natural resource managers. Differences in the interest of local communities and conservation program often lead to conflict between the two. Prior research has identified residents' attitudes toward tourism as an important factor in achieving successful sustainable tourism (Wang & Pfister, 2008; Látková & Vogt, 2012). Attitude of local communities toward conservation program is influenced by its impact on the livelihood strategies of people. In recent decades, wildlife or nature based tourism has been developed as a mitigating approach to address the economic difficulties of local communities. Benefits provided by the indirect use of natural resources especially from nature based and wildlife tourism are often used to generate opportunities for local communities and improve the tolerance of people toward presence of wildlife and conservation related actions (Ghate, 2003; Rastogi et al., 2015).

The main aim of development and promotion of tourism facilities in a particular area is to enhance the economic health and quality of life of the local communities along with conservation of the natural resources in the area. Studies on attitudes and perceptions of local people have contributed to the understanding of people's needs and aspirations and also in the identification of their opinions and suggestions regarding conservation issues. Many studies have emphasized that the attitude of local people is one of the main factors that affect the management and conservation effectiveness of the natural ecosystems. However, in densely populated areas where a large number of people are dependent on natural resources, effective conservation may not take place even when the people have a positive attitude toward conservation. As it generates livelihood opportunities for the local communities, wildlife tourism can be used as a tool to make people incline positively toward conservation and natural resource management programs. Communities are the most important stakeholders in the tourism industry as they are the ones who will be most affected by the decisions pertaining to tourism planning and development of their areas. Hence, understanding community attitudes are vital for the sustainability of rural tourism development (Eshliki & Kaboudi, 2012).

This study was conducted to assess the attitude of local communities of Nanda Devi Biosphere Reserve (NDBR) toward wildlife tourism in the area and to understand the role of monetary benefits of wildlife tourism in shaping the attitude of people in the area.

7.2 STUDY AREA

The NDBR is located in the State of Uttarakhand, India. It covers an area of 6020.4 km^2 at an altitude of 1800–7817 m in the Western Himalayas. The Reserve comprises of Nanda Devi National Park (NDNP) and Valley of Flowers National Park (VOF) interspersed with villages and cultivated lands (Fig. 7.1), which is termed as buffer zone of the National Parks. The buffer zone is surrounded by a multipurpose transition zone. Both the parks have been identified as Natural World Heritage Site (Bosak, 2008).

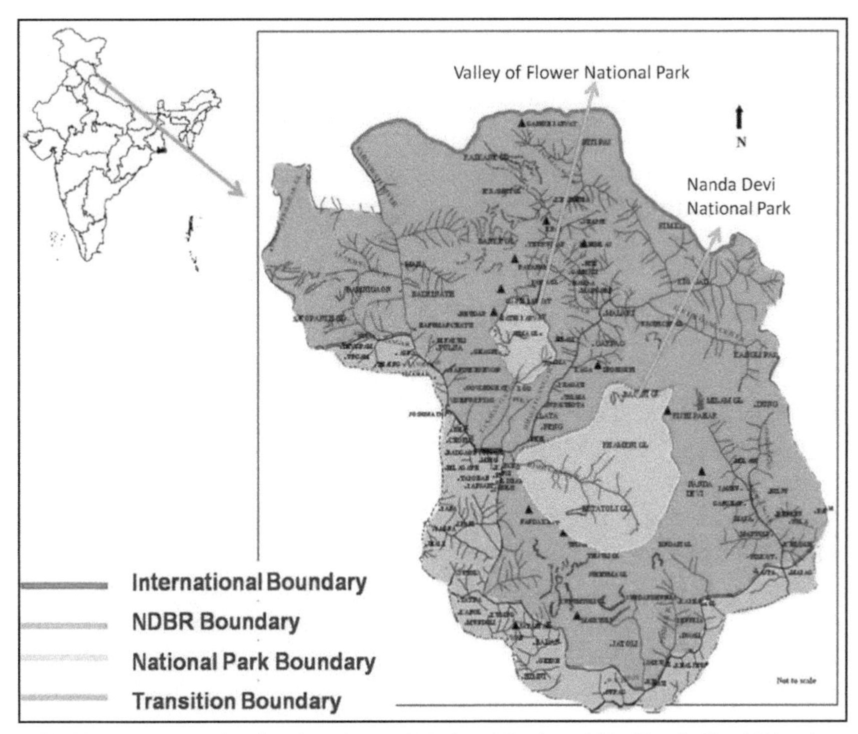

FIGURE 7.1 Map showing locations of National Parks within Nanda Devi Biosphere Reserve, India and different administrative zones.

Though no human habitation is present inside the National Parks, there are 47 villages located in the buffer zone and 33 villages in the transition zone. The Bhotia (Indo-Mongoloid) and the Garhwali (Indo-Aryan) are the main communities in the area. They are entirely dependent on natural resources for their sustenance and well-being. Large altitudinal variation and rich natural and cultural diversity make NDBR one of the hotspots of cultural, adventure, and nature tourism. The Reserve has been declared as World Heritage Site by UNESCO for its unique natural diversity (Bosak, 2008).

The Nanda Devi is the second highest peak of India, located inside the NDNP. It was once among the most popular destinations in the country for mountaineering till the ban on tourism in 1982 because of negative impacts of tourism on the landscape. Ban on tourism has affected the local economy and has raised conflict between the management and communities in the area. Nature based and cultural tourism has provided alternative livelihoods for local communities in the region by generating tourism related occupations and by creating market for local products. The presence of Hindu and Sikh shrine *Badrinath* and *Hemkunt* results in numerous visits to NDBR for religious purpose (~1.5 million per year). The religious tourists do not require permit and do not have to pay any entry fee, resulting in a high volume of unregulated tourists.

7.3 METHODOLOGY

The data on attitude of local communities toward wildlife tourism was collected in three stages through review of secondary information, selection of villages for extensive survey and semi-structured questionnaire based interviews (Harris & Brown, 2010; Badola et al., 2014). The first stage involved a rapid assessment of the study area in order to obtain an overall perspective. In the second stage, secondary data about all the villages of NDBR was gathered and information on type of and distance from nearby forest was collected. Hierarchical cluster analysis was carried out using secondary information—demographic parameters, development parameters, distance from forest, and condition of forest. A total of 11 clusters were formed and representative villages were selected for an intensive study. In the third stage, the primary data for selected villages was collected (Torgler & Garcia-Valiñas, 2005; Sodhi et al., 2010). Households (HH) in selected villages were sampled on a random basis. Head of the HH or any family member older than 18 years was interviewed (irrespective of gender) to get reliable information. A structured interview based questionnaire was developed to assess the attitude of local people toward wildlife tourism. To avoid personal and respondent biasness (Sheil & Wunder, 2002), the questionnaire was first tested in the field with 25 interviews and slight modifications were made to suit the conditions of the study area.

Individuals above the age of 18 years were approached and on confirming their willingness to participate in the questionnaire survey were interviewed. Initially ~780 villagers were approached, of which 16 refused to grant the interview; hence, the response rate was about 98%. Respondents were not offered any monetary or non-monetary benefits.

A total of 764 interviews were conducted in 22 selected villages. Personal information of respondents was collected. Respondents were asked to assign scores on the scale of 4–1 (four being the highest and one the lowest) (McIntyre et al., 2008; van Riper et al., 2012). Technical jargons were avoided while framing the questions to make them easy for the respondents to understand. The questions asked were in local language and related to the respondent's daily interaction with natural resources. Questions were asked indirectly to avoid the tendency to pick the first option (Sodhi et al., 2010). Around 200 HH were directly and indirectly dependent on tourism-related activities. One of the authors personally conducted all the interviews and was well versed in the local language of the area, therefore, did not require the assistance of any translators.

7.4 RESULTS

7.4.1 SOCIO-ECONOMIC PROFILE OF SURVEYED HOUSEHOLDS

A total of 764 HH were interviewed to assess the socio-economic condition of the local communities in NDBR. In that, ~44.6% and 55.4% of the respondents were men and women, respectively. In it, ~66% of the HH belonged to the Garhwali community and 33.9% to the semi-nomad transhumant Bhotia community. The average age of the respondents was 45.8 ± 0.55 years. The education level of the area is poor as only about 10% of the sampled population were graduates or had a higher degree. A lot of respondents had only primary education while 20.2% had secondary school education. Main occupation of the people residing in NDBR is agriculture followed by business (mainly related to tourism), government jobs, and private sector jobs (Fig. 7.2). Most of the people employed in private sector were working in the cities while few were employed in the hydroelectric projects in the area. The average annual income per HH was US$ 1,471.48 ± 51.27/hh/year which included income from agriculture, floriculture, horticulture, bee keeping, dairy, government jobs, and private sector jobs (Fig. 7.2). Natural resources contributed ~16% (US$ 235.62 ± 7.63/hh/year) to the total income of dependent HH. Average HH income for the HH (n = 200) involved in tourism was US$ 2,391.61 ± 232.06/year/hh with tourism contributing US$ 1,263.59 ± 156.44/year/hh. The income from tourism differs with type of occupation and directly involved HH earned more than HH indirectly involved. Although, the percent of contribution is higher for the indirectly involved economically weaker HH (Table 7.1). Indirectly involved HH generally sell their products to shops and middle men who then sell them to tourists.

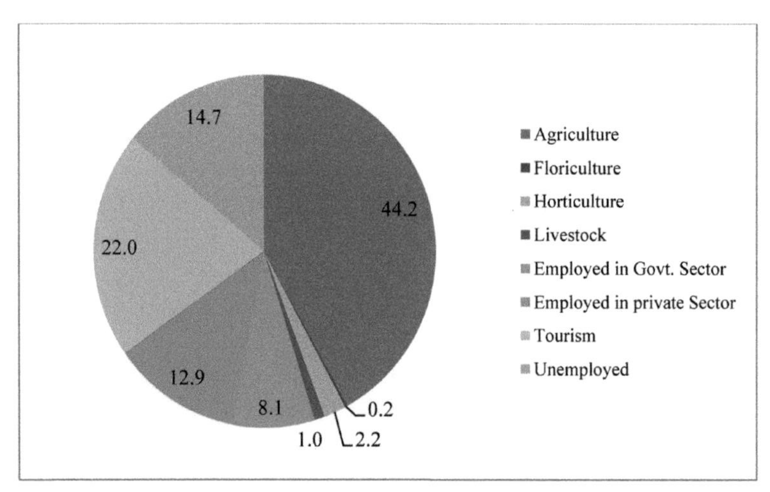

FIGURE 7.2 Occupational pattern (%) of members of households involved in tourism in Nanda Devi Biosphere Reserve, India.

TABLE 7.1 Contribution of Tourism to Total Annual Income of Households Involved in Tourism in Nanda Devi Biosphere Reserve, India.

Occupation	Income (US$/ hh/year)	Income from tourism (US$/ hh/year)	% of income from tourism
Tourist guide	1655.5 ± 160.5	132.4 ± 12.8	8
Homestay + Porter	2110.2 ± 286.3	316.5 ± 42.9	15
Work at Badrinath	1733.4 ± 115.4	1600.0 ± 106.8	92
Milk + Garland	2051.3 ± 136.9	242.7 ± 16.2	12
Mule	2088.3 ± 115.9	1253.0 ± 69.5	60
Photography	1266.0 ± 89.3	822.9 ± 58.0	65
Shop owners	3054.5 ± 202.7	2443.6 ± 162.2	80
Travel agency + tourist guide	6815.0 ± 1421.0	6338.0 ± 1321.5	93
Vehicle owner	2153.7 ± 190.8	1356.8 ± 120.2	63
Indirectly involved (including farmers, garland makers, herb collectors, and dairy workers)	806.2 ± 80.4	403.1 ± 40.2	50

7.4.2 ATTITUDE OF LOCAL COMMUNITIES TOWARD WILDLIFE TOURISM

About 47% HH representatives ($n = 275$; $N = 764$) regarded wildlife and nature based tourism as the most important service provided by the nearby natural areas while rest of the respondents did not place any value on tourism and recreation as it provides economic benefits to only few HH. Out of 275 HH, 12.4% valued the service as four while 20% valued it as three, 10.6% as two, and 3.9% as one (Fig. 7.3). In total ~3% of female and 37% of the male respondents scored the tourism as four and considered it as the most important ecosystem service while about 11% of female and 10% of male respondents scored the tourism as one (Fig. 7.4). There was a significant difference between the scores assigned by the female and male respondents. Most of the respondents (95%, $n = 764$) insisted that facilities should be promoted in and around their villages to facilitate and support the ongoing tourism related activities. All of the respondents said that they would like a government-supported program that provides training and vocational facilities for local people, mainly youth, to aid their involvement in organized nature based tourism activities.

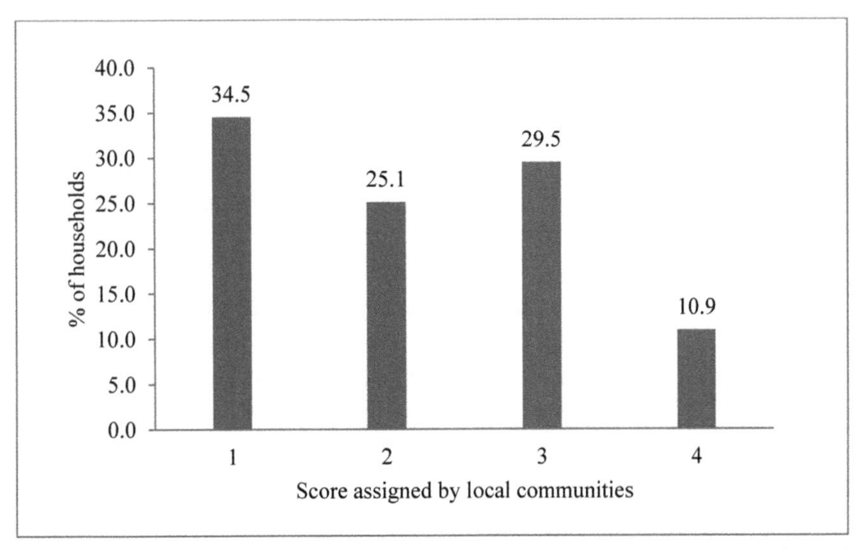

FIGURE 7.3 Score assigned by local communities to wildlife tourism in Nanda Devi Biosphere Reserve, India.

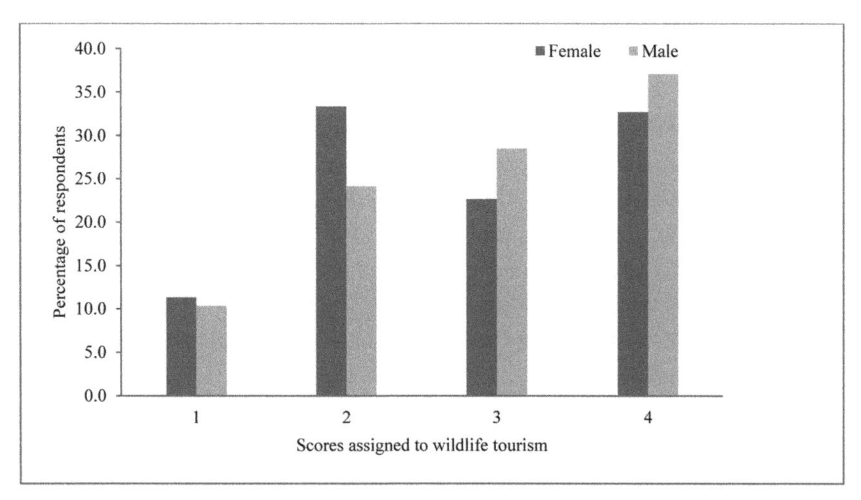

FIGURE 7.4 Difference in the score assigned by the male and female respondents to wildlife tourism in Nanda Devi Biosphere Reserve, India.

7.4.3 FACTORS INFLUENCING ATTITUDE TOWARD WILDLIFE TOURISM

The attitude toward wildlife and nature based tourism was related to the average age of a particular HH. HH with younger members (aged less than 30 years) and older members (aged more than 60 years) were found to be more positive toward nature based and wildlife tourism than HH with more elderly members. HH with more members in age category 15–30 years placed a higher value on tourism and scored it as four and three (~70% of the young members). The variation in the attitude of people with age is due to ability of younger people to participate in the tourism-related activities such as trekking, climbing, and so forth, which are difficult for the elderly people. The attitude of people (age >60 years) was also positive toward wildlife tourism, as they were involved in tourism before its ban in the NDBR region. Interestingly, people between 40 and 60 years of age were not very positive about wildlife tourism as they are not able to take part in present tourism-related activities due to its physical requirements and were not involved tourism before ban.

We observed that the years of schooling influenced the attitude of people toward wildlife based tourism in the area. The more educated respondents (mainly people who have at least spent 12 years in the school) assigned a higher value to wildlife tourism and scored it as four and three while the less educated people scored it as two and one. The well-educated and trained individuals have greater prospects in the tourism industry as they have the ability to interact with high end (tourists with more purchasing power and good educational background) and aware wildlife tourists and can partake in the physically strenuous tourism-related activities in the difficult terrain of the area. The less educated or illiterate members of the community often get involved in menial jobs and get fewer benefits. Being positive toward wildlife tourism implies the support for the conservation of wild animals in the area. Local communities repeatedly face loses due to wildlife, hence, do not support the conservation practices and are often negative toward wildlife tourism. HH with a higher dependence on natural resources and without a consistent income from other livelihood sources were found to be more positive toward tourism-related activities in the area. HH with income less than US$ 3000/annum wanted government and non-government agencies to develop the tourism infrastructure in their village and in nearby areas to support their economic health and reduce their vulnerability to change and loss of livelihoods and scored wildlife tourism as four. The scores assigned by HH reduced with the total HH income. Similarly, HH with more direct dependence on natural resources (dependence for provisioning services especially for fuel wood, fodder) were more positive toward the wildlife tourism and scored it as four and three while people employed in the public and private sector or with a regular income source are not much concerned about tourism promotion in the area.

7.5 DISCUSSION

The NDBR mainly hosts religious, mountaineering and expedition tourism and has potential as a sustainable wildlife tourism destination as it is home to many charismatic wild animals such as snow leopards, musk deer, and Himalayan black bear along with many rare species of birds, which provides great opportunities to visitors interested in wildlife. Nature based tourism often attracts tourist who can and are willing to pay more to enjoy the recreational services of any wild landscape and opt for sustainable home stays and local products. On the other hand, leisure and religious tourists come for a particular purpose and do not invest in exploring the scenic beauty of the area, providing livelihood to only few people by generating niche opportunities.

Results of this study show that the attitude of young members of the local communities in NDBR is significantly positive toward nature based and wildlife tourism. They placed high values on tourism as it provides them with the opportunity to earn more with relatively less physical and economic investment. Traditional occupations like agriculture and forest based occupations like livestock rearing are still the main occupations in the area and are highly vulnerable to change in local climatic conditions and introduction of international products to the local market. Not only does nature based tourism provide recreational opportunities to the visitors, it also provides a less vulnerable livelihood opportunity to the local people and makes them resilient to changes in their social and economic conditions. Hence, more young people are shifting to tourism based livelihoods in the area. Tourism provides monetary benefits very quickly as people involved get paid up front while traditional livelihoods often need more investment in terms of time, money, and physical work.

Local communities of NDBR wanted more investments from government and tourism-related organizations to promote sustainable nature based and wildlife tourism in the area. They also suggested that native species should be planted in the area and the local people should be trained as guides and travel agents, and should get tourism related equipment at subsidized rates. There was a difference in the attitude of the male and female members of the community as most of the male members are primarily engaged in market based occupations while the female members are involved in the collection of natural resources, agriculture, and livestock rearing and do not get any major benefit from tourism as such. Tourism is thought to benefit the local communities by providing alternate livelihood opportunities, but it is not always the case (Karanth & DeFries, 2011). A very insignificant proportion of the population residing close to PAs in India benefitted out of nature based tourism. In Chitwan National Park, Nepal, even though local employment was found to be high, it was the outsiders that largely garnered the financial benefits (Bookbinder et al., 1998; Spiteri & Nepal, 2008). Sharing benefits with the local communities and building support among private enterprises for conservation initiatives is a must for sustainable wildlife conservation (Karanth & Nepal, 2012). This will help in shaping a more positive attitude toward wildlife conservation.

Biodiversity conservation is crucial for the functioning of ecosystems and for continuous flow of their services (Hooper et al., 2005). The positive attitude of local communities, who share the services with the wild animals, is vital for the success of conservation programs. Tourism provides alternative and more resilient livelihood opportunities to the local people. Usang (1995) observed that negative attitudes of the people resulted in forest destruction. Understanding the attitudes of dependent community helps in identifying and realizing their ideas, opinions, and suggestions regarding conservation and management of natural resources (Badola et al., 2012). Although conservation via the ecosystem approach is determined by bigger policy decisions, the sustainable use of resources depends mainly on communities living in the vicinity of natural systems (Pyrovetsi & Daoutopoulos, 1997; Badola et al., 2012).

Prior to developing tourism resources, it is critical to understand the residents' opinion and attitudes to garner their support for future development (McGehee & Andereck, 2004). Long-term studies of local attitudes are also essential to fully identify the impacts of PA tourism. Tourism associated with economic, environmental and sociocultural benefits (Kuvan & Akan, 2005) must be viewed as a "community industry" (Murphy, 1985) wherein the residents are actively involved in the decision-making process pertaining to tourism development with the overall goal of improving the quality of life of the local communities (Fridgen, 1991; Látková & Vogt, 2012).

As per the study conducted by Walpole and Goodwin (2001) that examined the local attitudes toward PA tourism and the effects of tourism benefits on local support for Komodo National Park, Indonesia, while positive attitude toward tourism were directly related to the receipt of economic benefits and to support for conservation, there was no such relationship between receipt of tourism benefits and support for conservation, implying that benefits from PA conservation make no difference to local support for conservation. Distributional inequalities in tourism benefits, local inflation, and tourist dress code were common complaints.

Assessment of the attitude and behavior of local communities toward sustainable nature based and wildlife tourism can be used to understand the opportunities and threats that the conservation and management programs might face. These issues can be dealt by addressing the issues highlighted by the local communities and considering the value local people place on the flow of environmental services.

7.6 ACKNOWLEDGMENTS

This work has been carried out under the project "An integrated approach to reduce the vulnerability of local community to environmental degradation in the Western Himalayas," funded through the Grant-in-Aid funds of the Wildlife Institute of India (WII). We thank the Director and the Dean, WII for logistic and technical support. We express our gratitude to Prof. Mahmood Khan, Head, Department of Hotel, Restaurant, and Institutional Management at Virginia Polytechnic Institute and State University, Prof. Johra Fatima for their help, support, and encouragement.

KEYWORDS

- **wildlife tourism**
- **local communities**
- **questionnaire based interviews**
- **natural resource management**
- **Nanda Devi Biosphere Reserve**

REFERENCES

Badola, R.; Barthwal, S.; Hussain, S. A. Attitudes of Local Communities towards Conservation of Mangrove Forests: A Case Study from the East Coast of India. *Estuar. Coast. Shelf Sci.* **2012,** *96*, 188–196.

Badola, R.; Hussain, S. A.; Dobriyal, P.; Barthwal, S. An Integrated Approach to Reduce the Vulnerability of Local Communities to Environmental Degradation in Western Himalayas. Study Report, Wildlife Institute of India: Dehradun, India, 2014; p 300.

Ballantyne, R.; Packer, J.; Falk, J. Visitors' Learning for Environmental Sustainability: Testing Short-and Long-Term Impacts of Wildlife Tourism Experiences Using Structural Equation Modelling. *Tour. Manage*. **2011,** *32* (6), 1243–1252.

Bookbinder, M. P.; Dinerstein, E.; Rijal, A.; Cauley, H.; Rajouria, A. Ecotourism's Support of Biodiversity Conservation. *Conserv. Biol.* **1998,** *12* (6), 1399–1404.

Bosak, K. Nature, Conflict and Biodiversity Conservation in the Nanda Devi Biosphere Reserve. *Conserv. Soc*. **2008,** *6* (3), 211.

Curtin, S. Wildlife Tourism: The Intangible, Psychological Benefits of Human–Wildlife Encounters. *Curr. Issues Tour*. **2009,** *12* (5–6), 451–474.

de Pinho, J. R.; Grilo, C.; Boone, R. B.; Galvin, K. A.; Snodgrass, J. G. Influence of Aesthetic Appreciation of Wildlife Species on Attitudes towards their Conservation in Kenyan Agropastoralist Communities. *PLoS ONE.* **2014,** *9* (2), e88842.

Eraqi, M. I. Local Communities' Attitudes towards Impacts of Tourism Development in Egypt. *Tour. Anal.* **2007,** *12* (3), 191–200.

Eshliki, S. A.; Kaboudi, M. Perception of Community in Tourism Impacts and their Participation in Tourism Planning: Ramsar, Iran. *J. Asian Behav. Stud.* **2012,** *5* (2), 51–64.

Fridgen, J. D. *Dimensions of Tourism;* Educational Institute of the American Hotel & Motel Association Publishers: US, 1991.

Garrod, B.; Wornell, R.; Youell, R. Re-Conceptualising Rural Resources as Countryside Capital: The Case of Rural Tourism. *J. Rural Stud.* **2006,** *22* (1), 117–128.

Ghate, R. Global Gains at Local Costs: Imposing Protected Areas: Evidence from Central India. *Int. J. Sust. Dev. World Ecol.* **2003,** *10* (4), 377–389.

Harris, L. R.; Brown, G. T. L. Mixing Interview and Questionnaire Methods : Practical Problems in Aligning Data. *Pract. Assess Res. Eval.* **2010,** *15*, 1–19.

Hooper, D. U.; Chapin, F. S.; Ewel, J. J.; Hector, A.; Inchausti, P.; Lavorel, S.; Lawton, J. H.; Lodge, D. M.; Loreau, M.; Naeem, S.; Schmid, B.; Setälä, H.; Symstad, A. J.; Vandermeer, J.; Wardle, D. A. Effects of Biodiversity on Ecosystem Functioning: A Consensus of Current Knowledge. *Ecol. Monogr.* **2005,** *75*, 3– 35.

Karanth, K. K.; DeFries, R. Nature Based Tourism in Indian Protected Areas: New Challenges for Park Management. *Conserv. Lett.* **2011,** *4* (2), 137–149.

Karanth, K. K.; Nepal, S. K. Local Residents' Perception of Benefits and Losses from Protected Areas in India and Nepal. *Environ. Manage.* **2012,** *49* (2), 372–386.

Kuvan, Y.; Akan, P. Residents' Attitudes toward General and Forest-Related Impacts of Tourism: The Case of Belek, Antalya. *Tour. Manage*. **2005,** *26* (5), 691–706.

Látková, P.; Vogt, C. A. Residents' Attitudes toward Existing and Future Tourism Development in Rural Communities. *J. Travel Res*. **2012,** *51* (1), 50–67.

McGehee, N. G.; Andereck, K. L. Factors Predicting Rural Residents' Support of Tourism. *J. Travel Res.* **2004,** *43* (2), 188–200.

McIntyre, D.; Garshong, B.; Mtei, G.; Meheus, F.; Thiede, M.; Akazili, J.; Ally, M.; Aikins, M.; Mulligan, J. A.;Goudge, J. Beyond Fragmentation and towards Universal Coverage: Insights from Ghana, South Africa and the United Republic of Tanzania. *Bull. World Health Organ.* **2008,** *86* (11), 871–876.

Murphy, P. E. *Tourism: A Community Approach;* Methuen Publishers: New York, NY, 1985; p 200.

Pyrovetsi, M.; Daoutopoulos, G. Contrasts in Conservation Attitudes and Agricultural Practices between Farmers Operating in Wetlands and a Plain in Macedonia, Greece. *Environ. Conserv*. **1997,** *24* (01), 76–82.

Rastogi, A.; Hickey, G. M.; Anand, A.; Badola, R.; Hussain, S. A. Wildlife-tourism, Local Communities and Tiger Conservation: A Village-Level Study in Corbett Tiger Reserve, India. *Forest Policy Econ.* **2015,** *61,* 11–19.

Ritchie, J. B. Crafting a Destination Vision: Putting the Concept of Resident Responsive Tourism into Practice. *Tour. manage*. **1993,** *14* (5), 379–389.

Sandbrook, C. Local Economic Impact of Different Forms of Nature Based Tourism. *Conserv. Lett*. **2010,** *3* (1), 21–28.

Sekhar, U. N. Local People's Attitudes towards Conservation and Wildlife Tourism around Sariska Tiger Reserve, India. *J. Environ. Manage*. **2003,** *69* (4), 339–347.

Sheil, D.; Wunder, S. The Value of Tropical Forest to Local Communities: Complications, Caveats, and Cautions. *Conserv. Ecol.* **2002,** *6* (2), 9.

Sodhi, N. S.; Posa, M. R. C.; Lee, T. M.; Bickford, D.; Koh, L. P.; Brook, B. W. The State and Conservation of Southeast Asian Biodiversity. *Biodivers. Conserv*. **2010,** *19* (2), 317–328.

Spiteri, A.; Nepal, S. K. Distributing Conservation Incentives in the Buffer Zone of Chitwan National Park, Nepal. *Environ. Conserv*. **2008,** *35* (01), 76–86.

Torgler, B.; Garcia-Valiñas, M. A. *The Willingness to Pay for Preventing Environmental Damage;* (No. 2005–22). Center for Research in Economics, Management and the Arts (CREMA): Zürich, Switzerland, 2005.

Usang, E. N. Schools of Philosophy in Environmental Education. In *Philosophical Issues in Environmental Education;* Emeh, J. U., Ed.; Macmillan Nig. Publishers Ltd.: Ibadan, Nigeria, 1995.

van Riper, C. J.; Kyle, G. T.; Sutton, S. G.; Barnes, M.; Sherrouse, B. C. Mapping Outdoor Recreationists' Perceived Social Values for Ecosystem Services at Hinchinbrook Island National Park, Australia. *Appl. Geogr*. **2012,** *35* (1), 164–173.

Walpole, M. J.; Goodwin, H. J. Local Attitudes towards Conservation and Tourism around Komodo National Park, Indonesia. *Environ. Conserv*. **2001,** *28* (02), 160–166.

Wang, Y. A.; Pfister, R. E. Residents' Attitudes toward Tourism and Perceived Personal Benefits in a Rural Community. *J. Travel Res.* **2008,** *47*, 84–93.

CHAPTER 8

LOCAL RESIDENTS' ATTITUDE ON WILDLIFE TOURISM: SUPPORT, VIEWS, BARRIERS, AND PUBLIC ENGAGEMENT

NATALIE KING[1*] and VIKNESWARAN NAIR[2]

[1]*International Tourism and Hospitality Management, Taylor's University, Subang Jaya, Malaysia*

[2]*Sustainable Tourism Management, Taylor's University, Subang Jaya, Malaysia*

[*]*Corresponding author. E-mail: natoking@yahoo.com*

CONTENTS

ABSTRACT

In 2010, the number of domestic and international tourists who visited Sabah amounted to ~2.63 million. The Lower Kinabatangan area is theming with wildlife and encompasses a network of conservation areas that is home to some of Malaysia's native peoples, for example, Orang Sungai "River People." Although there are studies done worldwide that identify issues of wildlife value orientations there is a gap of in-depth probing for the beliefs on the value of wildlife as a community-based ecotourism product by the local community and tourist. The aims of this chapter are perceptions of local residents' toward wildlife value as a tourism product in a community-based setting.

> **"The value of wildlife is seen at several levels according to an individual's perception of its personal or group uses"**

The majority of local residents have shown valuable concerns for conservation of the ecosystem and community involvement in wildlife tourism ventures. The chosen case study has recognized that community-based ecotourism can be a means of habitat protection and income earnings for both wildlife and the local community once conducted within proper and manageable guidelines limiting tourists and developmental structures. More interestingly, the results show that the way wildlife is experienced may be leaning toward an increase in a non-consumptive value. This development may not only contribute to an increased satisfaction level of tourists but also be an attraction for new tourists searching for a participatory role and contributing by simply choosing a homestay accommodation.

8.1 INTRODUCTION

This chapter will encompass aspects of value covering basic knowledge of both non-consumptive and consumptive wildlife tourism (CWT) as a community-based ecotourism product. It will also seek perceptions of the local community on factors that contribute to wildlife tourism. The value of wildlife is seen at several levels according to an individuals' perception of its personal or group uses. In this case, value is researched from the local community point of view. The bases for this view are non-consumptive vs. consumptive where wildlife viewing, community-based ecotourism activities, and habitat loss are considered and discussed with the aim of acquiring views on non-consumptive use. However, wildlife viewing practices organized by the study site of Miso Walai Homestay program is also coupled with conservation efforts encouraging participation from the tourist.

Malaysia is located in South East Asia region with subtropical geographical features and with a population of 29.72 million (Government Census, 2015). The country is unique for its two separate land masses with the mainland peninsula (West Malaysia), and East Malaysia comprising the state of Sabah and Sarawak on the third largest island of the world, Borneo. In recent surveys Malaysia's tourists' arrival in 2014 amounted to 27.44 million with USD 12 billion in tourist spending. For the state of Sabah, 2.75 million tourists were recorded in 2014, which consist of the total arrivals for both international tourists at 29.7% and national tourists at 70.2% (Corporate Tourism Government, 2015). Sabah is situated at the northeast corner of Borneo, boasting a highly rich and diverse fauna and flora, and is one of the world's "hotspots of biodiversity." This diversity creates a wildlife tourism product that Sabah can offer to interested tourists coupled with rich indigenous heritage. Kinabatangan riverside areas are home to some of Borneo's most sacred fauna endemic to this region which include Orang-Utans (endangered on the International Union for Conservation of Nature (IUCN) 3.1 list), Agile Gibbon (endangered on the IUCN 3.1 list), Proboscis Monkey (endangered on the IUCN 3.1 list), Asian Elephants (endangered on the IUCN 3.1 list), Wild Otters (endangered on the IUCN 3.1 list), Borneo Sumatran Rhinoceros (Status: Critically endangered 3.1 Pop. trend: decreasing), and Hornbills and Crocodiles (Endangered United States Fish and Wildlife Service (EN-US FWS)) (IUCN Red List of Threatened Species, 2015).

Recently, Malaysia has incorporated Wildlife Conservation Act 2010, Sabah Biodiversity Enactment 2001, and Animal Act 1953 (Revised 2006) to assist in implementing stringent laws to be enforced. According to Higginbottom (2004), regions of high biodiversity of charismatic fauna represents 40–60% of international visitors, whereas "wildlife related" tourism accounts for 20–40%. With this in mind, the project "Heart of Borneo" was formed as a tri-country agreement amongst Malaysia, Indonesia, and Brunei Darussalam to incorporate some 220,000 km^2 of equatorial rainforests demarcated to ensure the effective management and conservation of protected areas within the area (Hamzah, 2004).

Some 27,000 hectares of The Lower Kinabatangan area was declared as a protected area in 1997 and later converted to a wildlife sanctuary in 2001. A protected area defined by the IUCN and World Wildlife Fund for Nature (WWF) is an area of land and/or sea especially dedicated to protecting and maintaining biological diversity associated with cultural resources managed through legal and other effective means (1999). The area is classified as high conservation value (HVC) which was used in the past for logging, human settlements, and plantation now mainly secondary forests of which efforts are being made to regenerate growth and provide forest cover for wildlife. According to Butler (2015), the percentage of forests cleared in Sabah between the years 2000 and 2005 was 3.9% and in 2010 this rate had doubled. According to WWF, during the 1980's oil palm began to replace other cash crops and has now become the dominant crop along the Kinabatangan River (WWF, 2009). To regain lands from oil palm plantations to its near natural state, the Kinabatangan Corridor of Life (KCoL) initiative was formed by WWF in 1997 focusing on providing a path suitable for easy movement of many species including the endangered pygmy elephant, proboscis monkey, orang-utan, rhinoceros, crocodiles, and hornbills.

The Miso Walai community in this region has captured the opportunity to work together to revitalize parts of the riverside which directly benefits the environment and community wildlife tourism product. Since implementation of WWF-Malaysia Strategy Sabah Plan (2012–2020) an approximate land mass of 1,174,398 hectares or about 15.95% of Sabah's land area was dedicated to for replanting activities (Institute of Development Studies Sabah, 1996, 2015). In such small-scale tourism activities, the benefit of addressing animal protection laws are necessary when considering tourist profiling as expressed by Manfredo (2008). He further added that attitudes toward wildlife management can be useful in planning for certain tourists' typology. Involving the community in this process assists in managing and controlling activities that disrupt or violates rules and regulations of protected areas. In terms of tourists, they may provide much needed revenue to sustain wildlife conservation activities as well as providing economic benefits to the community.

Community members recognized the need for conservation by acknowledging the seasonal natural changes such as dry and flood seasons as well as human induced changes such as river pollution, forest loss, and habitat loss. Their concerns brought insight into the past 20–30 years and stated that river pollution may also be due to human induced alterations to the natural environment. According to Sabah Tourism Borneo Post (2012), the decline in water quality is mainly due to pollution from palm oil mills and oil palm plantations located in the area and logging in the 1960s has aggressively aid in the conversion of floodplain forests to oil palm plantations in the 1980s. Even though this problem was noted, tourists commended the value of the community's heed to recover low-forested areas.

These activities coupled with the rapid change in land use have resulted in several serious ecological problems to the habitats and wildlife. It was also recognized that 85% of the Kinabatangan flood plain has been converted to agriculture land use and the remaining forested areas (15%) are not all in good condition owing to illegal logging and forest fires (Hai, 2001). It was noted that community members recognized these problems, thus, providing their opinions of possible solutions to regain wildlife by habitat reforestation, conservation, sustainable economic growth, education and awareness, and stakeholders' corporation. Due to major floods and forest fires the remaining evergreen swamps are extremely important as a water catchment area for biodiversity and local communities of Lower Kinabatangan. However, to an already damaged area due to unsustainable logging, plantation agriculture, mining, and hunting challenges are faced with the task of repurchasing land from plantation owners to accomplish the aims of the Heart of Borneo project KCoL.

8.2 MISO WALAI COMMUNITY TOURISM PRODUCT

The main product mix of the Miso Walai community is their homestay. Malaysia's homestay program has surpassed its target of 23% of occupancy rate (The Star News, 2012). To focus on the chosen location, the village of Batu Puteh Lower Kinabatangan in Eastern Sabah, an area known for its natural and cultural heritage, is discussed. The local ethnic group of Orang Sungai meaning "River People" (one of Malaysia's 32 ethnic groups) gained their name during the British colonial years. The community-based support initiative of Model Ecologically Sustainable Community Conservation and Tourism (MESCOT) was founded in 1997 by community members in the five Orangi Sungai villages of Batu Puteh-Mukim, Perpaduan, Paris, Singgah Mata, and Mengaris. Their aim is to develop tourism as a sustainable form of income generation through a bottom-up approach initiative which seeks to combine conservation and ecotourism activities confirmed by Mohd Hasim (2012, pers. Comm. 17 December).

The MESCOT program has achieved a step ahead in the restoration of lowland rainforest and has successfully constructed a zero waste and near-zero emissions eco camp, in addition to award winning homestay, cultural and wildlife tourism programs (Land Empowerment Animals People, 2011). The community utilized their own resources with the technical assistance from the main non-governmental organization (NGO) at that time WWF along with support from the government of Sabah. They then formed the tourism arm KOPEL Limited (Ltd) in 2003 to coordinate tourism and conservation activates. To date there are 35 family households participating in the Miso Walai Homestay program. For the year 2014, their earnings had increased to RM 181,000 from environmental restoration projects and RM 1.4 million with 5422 accommodation nights and 2810 total tourist numbers. This has been confirmed by Mohd Hasim (2015, pers. Comm. 03 August). Strong mention was also made of Wildlife Tourism as being the top three key factors that draw tourists to the community Source: World Wildlife Fund Malaysia (2009).

8.2.1 DEFINING WILDLIFE TOURISM

There are several angles to define wildlife one such definition was given by Rodger and Moore (2004) incorporating life of non-domesticated animals in the natural environment. According to Higginbottom (2004), wildlife tourism consist of two key variables which are commonly used to classify forms of wildlife tourism and the ability to maintain this form of tourism can provide a vital incentive for operators and/or host communities. They include non-consumptive wildlife tourism (NCWT) and CWT. Non-consumptive tourism involves recreational, activities that neither catch nor kill wild animals and *Consumptive tourism includes* when animals are being deliberately killed or removed, or having any of their body parts utilized or consumed. Non-consumptive wildlife value concept can be evaluated by utilizing the wildlife value orientation concept (WVO) which derives from value-attitude-behavior hierarchy (VAB) that states that values are abstractions from which attitudes and behaviors are manufactured (Manfredo, 2008). In this instance the community has viewed the value of wildlife as a form of NCWT thus its protection is seen as a benefit to their tourism product. There are aspects of reoccurring values of stakeholder involvement, environmental education and awareness, sustainability, and income noted in research conducted on Miso Walai community wildlife tourism product (King & Nair, 2012). These values will be discussed in the following sections.

8.3 STAKEHOLDER CORPORATION

In any corporation or business entity, stakeholders' involvement is as important as the product itself in this case community, government, tourists, and tourism businesses. As they say, many hands make magic but on a similar note many hands can also create situations where with many leaders can differ the cumulative goal. In situations where communities are directly affected, their presence should be slightly above the rest from initial stages of the decision-making process. On the beneficial side as noted by many community members, local government such as Sabah Forest Department and Sabah Wildlife Department can provide support by promoting conservation, providing jobs for locals as forest warden and policy enforcement. As noted by Buckley (2003), there are many systems and indicators that are inadequate to manage tourism impacts because ecological and management indicators are not feasible and therefore impracticable to managers. Thus, making joint partnerships between tourism entities and environmentalists are necessary in developing visitor management techniques. Studies conducted by Marion (2006) on the effects of ecological recreation on backcountry sites such as campsites and trails, indicated several negative impacts on biodiversity. Liaising with tour operators can help to promote tourism and possibly increase tourist visitation to community-based tourism programs. With the emergence of conservation and small-scale tourism activities it is possible that benefits can be reaped without consumptive usage of natural resources.

8.4 ENVIRONMENTAL EDUCATION AND AWARENESS

Environmental education is usually communicated to the public through social media and the education system. Community members are of the impression that visitor education should be designed to persuade visitors to adopt low-impact practices appropriately. It is believed that once visitors are aware of damaging effects of their actions they would modify their behavior to be more responsible. There is usually a concern about wildlife protection between tourists and community members in small-scale tourism entities. Views on the "use" of animals by humans can vary from regions, countries, cultures, traditions, or persons. According to Shani and Pizam (2008) since the influential book "Animal Liberation" was published in 1975 by the Australian philosopher Peter Singer, public concern over ethical treatment of animals has increased dramatically. If the current economic and social trend continues a sustained erosion of traditional orientations toward wildlife is likely the value orientations will shift from materialist to post-materialist values as explained by Fulton et al. (1996). It is mentioned that people who have better jobs, more education, and higher incomes in many western countries are more aware of animal rights laws, which will have its own impacts on the management strategies in the tourism industry.

8.4.1 COMMUNITY CAPACITY BUILDING

Value is also important in researching and understanding the visitors' purpose and required experience, aiding in tourist typology for future planning and management. One such example is through actionable research, capacity building, and environmental awareness in the case of wildlife protection in India (Sarkar & George, 2010). Similar capacity building concept was also utilized by the MESCOT initiatives in Kinabatangan. In respect to the community, environmental awareness can be transferred through Community Capacity Building (CCB) whereby initiatives are developed through a learning process with the guidance of a stakeholder and in the Miso Walai case, WWF was the facilitator of CCB. This concept was adopted by the United Nations Development Program (UNDP) in 1998, identifying three areas of significance—community, organizational, and individual levels. Their study further revealed that the key success of community-based tourism in the context of Miso Walai Homestay in Kinabatangan was due to community empowerment, planning, awareness, knowledge, and skills throughout the stages of development. The community was able to work with WWF in extensive researches that lead to the gazettement of Lower Kinabatangan area in 1997.

8.4.2 SUSTAINABILITY

Population and economic growth tend to persuade communities dwelling in a sustainable life to divert to activities that may not be in the realm of sustainability. However, members of Miso Walai community have embraced this concept with the notion that future generations will also have the benefit of these natural resources. Respondents from the local community incorporated the notion of sustainability by indicating their ideas of conserving the ecosystem for future generations and economic growth. It is important to maintain the four pillars of sustainable development (economic, environment, culture, and social) for community-based ecotourism to be successful.

Research conducted by Kruger (2005) included 188 case studies on habitat transformation that were analyzed from Africa, Central America, South America, and Asia. He made references of Malaysia pointing out that large-scale habitat transformation has been taking place to enhance the ecotourism experience. Nonetheless, sustainable case studies were seen to have conservation projects, local community's involvement, flagship species, non-consumptive use of wildlife, and also effective planning and management. It is credible to say that ecotourism can have a positive impact on both the local community and the environment. However, some entities use this as a marketing tool and possibly misusing the term to gain positive leverage in the eyes of the public (Woods, 2002). For sustainability to be successful it is important to set strategies with the involvement of key stakeholders not only in the present planning but also for its longevity.

Government conservation and sustainable economic growth initiatives are seen as plausible in the move toward a sustainable future for generations to come, however, management and policy enforcement at ground and local levels are still being questioned (India Info Line, 2012, 2015). Initiatives at the local level should also create a sense of ownership throughout societal values, therefore, acknowledging nature as valuable even if not used by humans at present time but its untouched value to non-human animals and for future generations. The Miso Walai community expressed the above value systems whereby reflecting concerns over the natural environment and biodiversity. The notion of "care" for non-human life and the environment can be linked to tourism as functionally and symbolically equivalent to other institutions that humans use to embellish and add meaning/value to their lives (Sharpley & Jepson, 2011). For some, a spiritual connection to natural places may be preconceived or felt at the point of experiencing nature itself. Knowing the value orientation of a visitor will enable tourism entities to understand the purpose of visiting and their required experience which can also assist in noting tourist typologies, awareness, and aid in planning and management.

8.4.3 PROTECTED AREAS AS A TOOL FOR SUSTAINABILITY

As population growth surges and economic persuasion increases, governments worldwide joined in the development of memorandum of understandings (MOUs) between countries such as Bulgaria, Romania, Guyana, Norway, Netherlands, and Denmark, under Article 6 of the Kyoto Protocol with the aim of combating global climate change, environmental degradation, and protection of indigenous communities (Mitchell, 2012). Endangered species are understood by respondents as few in populations, has less habitat, need special protection and there is the possibility of becoming extinct in a short period of time. One such known MOU is in Guyana where 371,000 hectares of forest was donated to the world in 2009 (Iwokrama Rainforest Research Centre, 2012, 2015). Similarly, Lower Kinabatangan was also given as a gift to the world in 2002 when the State Government (Sabah) declared the KCoL as a "Gift to The Earth" (Majail & Webber, 2006).

The rate of biodiversity loss will not only affect pockets or specific species but the ecosystem as a whole, including humans. Recent studies conducted by 24 scientists utilizing both empirical and historical data show that findings demonstrate that top predators in the food chain are enormous influencers of the structure, function, and biodiversity of most natural ecosystems. Sustainable use of parks is as important as dialogue with indigenous communities in an attempt to avoid conflict with their livelihood and tradition. This is the constant difficulty faced by indigenous communities during government decisions when allocating land for commercial use and even as protected areas and national parks. Previously many countries adopted a North American model of protected area designation, which resulted in the removal of indigenous population from within the protected area boundaries (Egales et al., 2000). As noted by Kruger (2005), in almost 40% of cases, the consequent involvement of local communities during planning, decision making, or as a substantial labor source, made the ecotourism projects and ventures sustainable through reduced need to practice consumptive land use. However, in recent years, parks have adopted the function of protecting the environment and local communities within and some even around the demarcated protected area.

8.4.4 INCOME

In terms of economic valuation, contingent valuation method (CVM) can be useful in the expression of willingness to pay (WTP) for an environmental service and measure of satisfaction. According to the definition and explanation given by Shani and Pizam (2008):

"WTP is the amount of money that a person is willing and able to pay to enjoy recreational facilities. It measures whether an individual is willing to forego their income in order to obtain more goods and services, and is typically used for nonmarket goods."

There are two main types of non-use value services that encompass forms of recreational use; these are existence and option value. Existence value is the enjoyment of biodiversity which involves use value—that is, it derives from a physical encountering with various plants and animals—satisfaction is also derived from simply recognizing that these forms of nature exist (Goulder & Kennedy, 2009). This concept also surrounds the notion of culture, reflects spiritual, diversity, and power of nature. The second approach noted by Goulder and Kennedy (2009), as the "option value" which refers to a higher value that people are willing to pay to preserve an environmental amenity, over and above the mean value (or expected value) of the use values anticipated from the amenity.

Interestingly, all respondents indicated that income through wildlife tourism benefits the community and their conservation efforts. Wildlife is seen as a unique selling preposition (USP) which combines a mix of services and products that is used not only to generate income for operational purposes but also to assist in conservation. As indicated by Butler and Hinch (1996), that communities who engage in wildlife tourism have found non-consumptive ways to utilize and co-exist with the ecosystem. Studies have shown that tourists venture into rural and nature areas to experience a sedated life away from modern life as a refuge from the pressures of modern life (Woods, 2011). Supporting the uniqueness of the product and the WTP to conserve species for future generations can be a positive outlook on NCWT. This links to an aspect of wildlife tourism which states that one of the requirements for sustainable use of wild species, for tourism or other purposes, is that there is a positive economic incentive for people living near such populations to conserve (Higginbottom, 2004). However, some visitors mentioned the difficulty to locate such an authentic product, which can be a benefit to secure the small-scale nature of the product adding the authenticity to the brand.

8.5 LIMITATIONS

Acknowledging limitations of this chapter is the study on a single wildlife tourism destination, therefore, quantitative and qualitative can be carried out which may probably yield more interesting results. It is also recommended that a wider and longer study over a period of years utilizing both quantitative and qualitative methods can yield a more comprehensive and statistical analysis.

8.6 CONCLUSION

There are communities that embrace the wildlife concept with positive attitude toward futuristic benefits of wildlife and more so the concept of wildlife tourism to gain economic benefits for current and future generations. Promoting wildlife tourism has been a major and deliberate conservation strategy in many less developed countries where poaching or persecution of wildlife as pests is a major threat to wildlife, particularly in Eastern Africa, where there is an informal policy of "wildlife pays, wildlife stays" (Higginbottom, 2004). The Miso Walai Community tourism program has found a way to combine both aspects of wildlife protection and economic growth. Research has highlighted four main ideas of community members' attitude toward incorporating fauna as a primary entity; income, stakeholder corporation, sustainability, and environmental awareness. In several cases tourism is seen as an economic benefit to the local community and to the environment through conservation revenue and local support. Tourist revenue is seen as a positive initiative to secure wildlife for future generations and once the revenue entity (in this case tourist) has experienced the positive outcome of their fiscal support, the WTP will be extended in support of the community.

In terms of sustainability, small-scale operations focusing on product authenticity is conveyed by community members by acknowledging all four pillars of sustainability (economic, ecology, culture, and politics). In their own expressions they indicated knowledge and accepted wildlife as a necessary combination for the success of their tourism venture. As indicated, knowledge of the environment plays a key role in policy making and vital research in supporting the need for protected areas, it also gives the community a wealth of confidence in the product itself. This authenticity and value is what drives the communities' "will" to succeed and the tourists' "will" to engage in community-based ecotourism activities. It also contributes greatly to the conservation and awareness offered to tourists and surrounding communities. The community has gained a valid understanding of reasons necessary to sustain resources for future generations and ways of utilizing wildlife in a non-consumptive and sustainable manner. In the advent of such models, working knowledge can be obtained for future community-based ecotourism programs.

The community's program not only offers visitors, volunteers, and researchers an opportunity to experience a wealth of fauna, flora, and cultural values but also enables them to get directly or indirectly involved through activities organized by the community. In this research it was found that tourists are engaged in conservation activities such as tree planting, environmental awareness talks, species behavioral patterns, cultural exchange, and understanding the way of life of the local community. This authenticity and value is what drives the communities' "will" to succeed and the tourists' "will" to engage in community-based ecotourism activities.

KEYWORDS

- **Kinabatangan**
- **Sabah**
- **wildlife tourism**
- **Miso Walai Homestay**
- **wildlife value orientation**
- **rural**

REFERENCES

Buckley, R. Ecological Indicators of Tourist Impacts in Parks. *J. Ecotour.* **2003,** *2* (1), 54–66. doi: 10.1080/14724040308668133.

Butler, R. Facts on Borneo. 2015. http://www.mongabay.com/borneo.html (accessed on August 11, 2015).

Butler, R.; Hinch, T. *Tourism and Indigenous Peoples;* International Thomson Business Press: India, 1996.

Corporate Tourism Government 2015; Tourist Arrivals and Receipts to Malaysia. http://corporate.tourism.gov.my/research.asp?page=facts_figures (accessed August 11, 2015).

Egales, P. F. J.; McCool, S. F. *Tourism in National Parks and Protected Areas: Planning and Management;* 1st ed.; CABI Publishing: Oxfordshire, UK, 2000.

Fulton, D.; Manfredo, M.; Lipscomb, J. Wildlife Value Orientations: A Conceptual and Measurement Approach. *Hum. Dimens. Wildl.* **1996,** *1* (2), 24–47.

Goulder, L. H.; Kennedy, D. *Interpreting and Estimating the Value of Ecosystem Services,* (April), 3–6, 2009.

Government Census 2015. https://www.statistics.gov.my/index.php?r=column/cone&menu_id=ZmVrN2FoYnBvZE05T1AzK0RLcEtiZz09. (accessed on August 23, 2015).

Hai, T. C.; Ng, A.; Prudente, C.; Pang, C.; Tek, J.; Yee, C. In *Balancing the Need for Sustainable Oil Palm Development and Conservation: The Lower Kinabatangan Floodplains Experience*, ISP National Seminar 2001, Strategic Directions for the Sustainability of the Oil Palm Industry Kota Kinabalu: Sabah, Malaysia, Jun11–12, 2001.

Hamzah, A. *Policy and Planning of the Tourism Industry in Malaysia,* Proceedings of the 6th ADRF General Meeting, Bangkok, Thailand, 2004.

Higginbottom, K. Wildlife Tourism. *CRC Sustain. Tour.* **2004,** *1* (1), 1–301.

India Info Line 2012; WWF Urges Investors to Promote Sustainable Palm Oil. www.indiainfoline.com. (Last Accessed on November 13, 2012).

India Info Line 2015; WWF Urges Investors to Promote Sustainable Palm Oil. www.indiainfoline.com (accessed on August 15, 2015).

Institute of Development Studies Sabah (IDS) 2015; The Sabah Tourism Master Plan. http://www.townplanning.sabah.gov.my/iczm/reports (accessed on August 8, 2015).

Institute of Development Studies Sabah (IDS) 1996; The Sabah Tourism Master Plan. http://www.townplanning.sabah.gov.my/iczm/reports. (Last Accessed on August 8, 2012).

Iwokrama Rainforest Research Centre 2012; http://www.iwokrama.org/wp (Last Accessed on October 20, 2012).

Iwokrama Rainforest Research Centre 2015; http://www.iwokrama.org/wp (accessed on July 20, 2015).

IUCN Red List of Threatened Species 2015; http://www.iucnredlist.org/ (accessed on July 22, 2015).

Krüger, O. The Role of Ecotourism in Conservation: Panacea or Pandora's Box? *Biodivers. Conserv.* **2005,** *14* (3), 579–600. doi:10.1007/s10531-004-3917-4.

Land Empowerment Animal People. New Sabah Times: Rosli Takes a Turn for the Better. http://leapspiral.wordpress.com/2011/11/18/rosli-takes-a-turn-for-the-better. (accessed on July 24, 2015).

Majail, J.; Webber, A. *Human Dimension in Conservation Works in the Lower Kinabatangan: Sharing PFW's Experience,* Proceedings of the 4th Sabah-Sarawak Environmental Convention, Malaysia, 2006.
Manfredo, M. Values, Ideology, and Value Orientations. In *Who Cares About Wildlife?* Springer: NewYork, NY, 2008; pp 141–166.
Marion, J. L. *Recreation Ecology Research in the Americas,* Proceedings of the Third International Conference on Monitoring and Management of Visitor Flows in Recreational and Protected Areas, University of Applied Sciences, Rapperswil, Switzerland, Sept13–17, 2006.
Mitchell, R. International Environmental Agreements Database Project (Version 2012.1). http://iea.uoregon.edu (accessed on August 7, 2012).
King, N.; Nair, V. The Value Orientation of Wildlife as a Community Based Ecotourism Product: A Case Study of Lower Kinabatangan. Sabah, Master's Thesis Dissertation, Taylors University, Malaysia, 2012.
Rodger, K.; Moore, S. A. Bringing Science to Wildlife Tourism: The Influence of Managers' and Scientists' Perceptions. *J. Ecotour.* **2004,** *3* (1), 1–19.
Sabah Tourism Borneo Post 2015;.http://www.sabahtourism.com/sabah-malaysianborneo/en/news/8620-sabah-expects-275-million-tourist-arrivals-masidi/. (accessed on August 3, 2015).
Sarkar, S. K.; George, B. P. Peace through Alternative Tourism: Case Studies from. *Tourism.* **2010,** *1* (1), 27–41.
Shani, A.; Pizam, A. Towards an Ethical Framework for Animal-Based Attractions. *Int. J. Contemp. Hosp. Manage.* **2008,** *20* (6), 679–693. doi:10.1108/09596110810892236.
Sharpley, R.; Jepson, D. Rural Tourism. *Ann. Tour. Res.* **2011,** *38* (1), 52–71. Available at: http://linkinghub.elsevier.com/retrieve/pii/S0160738310000678
Shen, S. X.; Zinn, H., Wang, A. Y. Assessing Wildlife Value Orientations in China: An Exploration of the Concepts and Methodology. *Recreation.* **2006,** *38* (1), 470–471.
The Star News 2012; Tourism to Give Malaysia a Boost. http://thestar.com.my/news/story.asp?file=/2012/5/2/nation /11192349&sec=nation (accessed on July 23, 2012).
Wood, M. E. *Ecotourism: Principles, Practice & Policies for Sustainability;* United Nations Environment Programme: France, Vol. 1, 2002.
Woods, M. *Rural*; Holloway, S., Valentine, G., Eds.; 1st ed.; Routledge: UK, 2011; p 336.
WWF 2009; Kinabatangan Corridor of Life. *Corridor Newsletter*. Issue 1, 2009.

CHAPTER 9

CONTRIBUTION OF WILDLIFE TOURISM TO TIGER RESERVES IN INDIA

BITAPI C. SINHA*

Protected Area Network, Wildlife Management and Conservation Education Department, Wildlife Institute of India, Dehradun, Uttarakhand, India

*E-mail: bcs_wii@yahoo.co.uk

CONTENTS

ABSTRACT

Tiger reserves in India are today important tourist destination and attract a large number of visitors, thus presenting a range of financial and non-financial opportunity for contribution and also challenges for management. This chapter presents an overview of the benefits of wildlife tourism to the park and the community living around. The financial contribution is arising from gate receipts, which includes the entry fees for vehicles and visitors, is deposited to the state exchequer. This chapter reveals how a portion of the revenue generated, by Tiger Reserve in the state of Madhya Pradesh, is shared with the community living around. This might not provide basics for food and clothing but complements infrastructure development like roads, schools, shelter, and water. We also examined how tourism is contributing to conservation by securing a sustainable livelihood for the community and creating awareness that results in an appreciation of and a positive attitude toward conservation of resources.

> **"Studies have indicated that wildlife tourism can create incentives for conservation..."**

9.1 INTRODUCTION

Worldwide tourism has increased manifold in protected areas (PAs) as it offers a wilderness-centric nature appreciation, experience largely to urban visitors (Eagles, 2007). Wildlife tourism, like any other consumer goods has continued to expand as a consequence of increasing wealth, interest, outgoing attitude, and improvement in transport (Sezgin & Yolal, 2012; Clough, 2013). Tourism is believed to provide benefits to PAs and nearby communities, who are dependent on forest resources, if planned carefully (Eagles et al., 2002; Tosun, 1999, 2000). Studies have indicated that wildlife tourism can create incentives for conservation (Jamal & Stronza, 2009), be a driver for economic change and empowerment that can make local communities less dependent on natural resources (Billgren & Holmen, 2008) and also create awareness amongst society (Karanth et al., 2012). The real value of tourism to conservation is much debated (Adams & Infield, 2003; Archabald & Naughton-Treves, 2002; Bloom, 2000; Cleary, 2006; Garnett et al., 2003; Lerner et al., 2005; Salzer & Salafsky, 2006; Sandbrook, 2008; Nagothu, 2003). Value of tourism in

PAs to conservation also depends on how much revenue flows to the local people whose lives are most affected by conservation actions (Archabald & Naughton-Treves, 2002; Sandbrook, 2008; Nagothu, 2003).

9.2 TIGER RESERVES IN INDIA

In 1970, there was concern about Indian Tigers *Panthera tigris* that lead to the establishment of Project Tiger in 1972 with the declaration of nine Tiger Reserves with the objective to "Preserve for all times, the area of such biological importance as a national heritage for the benefit of education and enjoyment of people." In 2005, a Tiger Task Force was set up, when Sariska Tiger Reserve in Rajasthan lost all its tigers. The Tiger Task Force along with other recommendations suggested the setting up of the National Tiger Conservation Authority (NTCA) to provide legal sanctity to the directives of Project Tiger and also address livelihood issues of local community surrounding the Tiger Reserves. The number of Tiger Reserves in India has increased from 9 in 1973 to 48 Tiger Reserves covering an area of 70,659.413 km^2 and a population of 2226 tigers, which is 70% of the world tiger population (Jhala et al., 2015).

9.3 GROWTH OF TOURISM IN TIGER RESERVES

Traditionally forest areas in India were renowned for sport hunting of wildlife by the Royals. Gradually the biodiversity-rich areas were declared as PAs with the mandate of conservation of natural heritage. Although the primary objective of PAs in India is strict protection of natural conditions but some amount of tourism and recreational activity is allowed for eight months, October–June, in a year. As per NTCA Guidelines, in the Tiger Reserve in India tourism is only permitted up to 20% of the tiger habitat as a tourism zone. In 2003, guidelines were issued for calculating the carrying capacity of the reserve for tourism. Each Tiger Reserve has calculated the carrying capacity as per the guidelines and this is strictly followed. Only the calculated numbers of vehicles are allowed with visitors on pre-identified routes.

Large mammals are often focal species for conservation as well as an attraction for tourism (Okello et al., 2008; Beschta & Ripple, 2009; Lyngdoh et al., 2014b; Winterbach et al., 2013). In the South Asian context, there is probably no greater a flagship species for conservation than the tiger (Goodwin, 1996; Narain et al., 2005). Tiger is an umbrella species used for conservation promotion (Goodwin, 1996; Leader-Williams & Dublin, 2000; TTF, 2005) and the Tiger Reserves attract a large number of visitors, predominantly domestic (Narain et al., 2005; Hannam, 2005; Karanth & DeFries, 2011). In India, public interest in tigers alone (apart from limited religious tourism that exists inside such PAs) has generated huge revenues, indicating the potential of financial gains (Narain et al., 2005) (Table 9.1).

TABLE 9.1 Visitors to Tiger Reserve.

S. No.	Tiger reserve	Year of declaration	Area (km²)	Number of visitors 2013–2014	Revenue in million US$
1.	Dudhwa	1987–1988	2201.7748	19,832	
2.	Corbett	1973–1974	1288.31	211,675	1.24
3.	Ranthambore	1973–1974	1411.291	324,325	0.54
4.	Sariska	1978–1979	1213.342	32,361	0.12
5.	Tadoba Andhari	1993–1994	1727.5911	103,616	
6.	Bandhavgarh	1993–1994	1598.1	113,940	
7.	Pench (MP)	1992–1993	1179.63225	54,027	
8.	Nagarhole	2008–2009	1205.76	67,277	0.45
9.	Dandeli-Anshi	2008–2009	1097.514	37,504	
10.	Periyar	1978–1979	925	780,853	0.78
11.	Parambikulam	2008–2009	643.662	51,726	
12.	Anamalai	2008–2009	1479.87	583,244	
13.	Kaziranga	2008–2009	1173.58	128,435	0.44
14.	Manas	1973–1974	3150.92	35,500	
15.	Sunderbans	1973–1974	2584.89	157,757	0.23
16.	Panna	1994–1995	1578.55	19,419	0.11
17.	Kanha	1973–1974	2051.791	133,671	1.03

The conversion rate of US$ 1 = Rupees 60 was used. Visitor data does not include pilgrims.

From the 1970s to early 1990s tourism in Tiger Reserves was limited with few enthusiastic visitors who would like to go into the forests and enjoy wildlife. The increase in media attention to wildlife, especially tigers, through cable Television especially National Geographic and Discovery channels, increase in earnings, development of transport facilities, the government of India providing leave travel concession to its employees, and their dependents, has led to domestic wildlife tourism becoming mass tourism. Tiger reserves are now preferred destination for visitors with many now deal with visitors who invest large amounts of money, time, and effort to experience these areas (Eagles, 1992; Karanth & DeFries, 2011; Narain et al., 2005).

The analysis of the volume of tourists in 2013–2014 presents a range of visitors from 19,419 in Panna Tiger Reserve to 780,853 in Periyar Tiger Reserve. Some of the Tiger Reserves like Corbett, Ranthambore, Tadoba, Bandhavgarh, Periyar, Annamalai, Kaziranga, the Sundarbans, and Kanha are popular visitor destinations where visitors throng to see a tiger from close distance in the wild. This can be attributed to better sightings of wildlife, easy accessibility, publicity and better, and reasonably priced visitor facilities. Tiger reserves like Sariska, Periyar, and Ranthambore have shrines located within, which are visited by pilgrims and some Tiger Reserves are affected by extremism because of which they do not receive visitors.

9.4 FINANCIAL CONTRIBUTION OF TOURISM

Today the trend is to use tourism as an economic development tool, with receipts from tourism forming a significant source of foreign exchange and visitor expenditures not only providing employment to the residents but also helping fund conservation of wildlife and biodiversity (McCool, 2009). Global analysis of the contribution of tourism revenue to threatened species conservation has shown that tourism revenues generated by parks protect significant proportions of global populations for a number of threatened vertebrates (Buckley et al., 2012; Morrison et al., 2012; Steven, 2011).

PAs are no longer managed in isolation and adopt stakeholder inclusive strategies. Community participation is regarded as an important tool for successful tourism development. However, in developing countries such participation is difficult to put into practice due to structural, cultural, and operational limitations in the tourism development process. While local communities are supportive of wildlife tourism in Tiger Reserves (Karanth & Nepal, 2012; Nagothu, 2003), however, tourism in PAs should lead to economic benefit to the local community and residents such that they have incentives to support conservation of the PAs (Bushell & McCool, 2007). Deriving tangible benefits from tourism activities within PAs can result in attitude to conservation among local communities (Liu et al., 2012; Snyman, 2012a). Ecotourism is thought to encourage local guardianship of biological resources by creating economic incentives for impoverished villagers or their communities (Bookbinder et al., 1998). It is assumed that wildlife tourism can benefit local communities with the jobs it creates (Cook et al., 1992).

9.5 PLOWING BACK OF TOURISM REVENUE

PA budgets in India do not cater for tourism. For the majority of the PAs in the country, the revenue generated from tourism, mainly entrance fee, vehicle fee, camera fee, and accommodation fee, goes to the government exchequer. Besides, the revenue generated by the tourism operators, hoteliers, and transporters is also not captured at the local level as the majority of the operators are from outside the area. In India, Madhya Pradesh is the only state where the revenue generated from tourism in the PAs under its jurisdiction is plowed back for development of visitor-related facilities, conservation related activities like development of habitat, and staff welfare and community need for relocated villages, forest villages, and revenue villages around the PAs. In Periyar Tiger Reserve, Kerala, 56% of the revenue generated from tourism is being given to Periyar Foundation to support ecodevelopment and people's participation (Karanth & DeFries, 2011; Periyar Foundation, 2007).

9.6 CASE STUDY FROM KANHA TIGER RESERVE

9.6.1 LOCAL ECONOMIC BENEFITS FROM TOURISM

Kanha National Park is one of the first nine areas declared as Tiger Reserve, in 1973. The tourism boom started in 1983–1984 as a sequel to the visit of Prince Phillip, the Duke of Edinburgh, and the late Mrs. Indira Gandhi, Prime Minister of India. A Large number of Indian and foreign tourists visit the park, and their numbers have increased from 38,236 visitors in 1998–1999 to 137,644 in 2014–2015, with a corresponding increase in revenue (Table 9.2).

TABLE 9.2 Monetary Benefits Derived by Kanha Tiger Reserve from Tourism.

Year	Total number of visitors	Revenue from tourism (US$)
2008–2009	137,295	639,354.00
2009–2010	154,024	508,431.00
2010–2011	174,773	833,437.00
2011–2012	182,354	997,513.00
2012–2013	130,457	735,085.00
2013–2014	133,671	1,035,157.00
2014–2015	137,644	1,087,781.00

Source: *Tiger reserve records, Madhya Pradesh forest department.*

The revenue generated is from the gate receipts, which includes the entry fees for vehicles and visitors, and it remains with the park in the form of a fund called the Vikas Nidhi.

9.6.2 REVENUE FROM WILDLIFE TOURISM SHARED WITH COMMUNITY

Eco-development Committees have been set up in the villages located in the proximity of the park. The eco-development committee comprises of adult villager above the age of 18 years. This is formed by the villagers in consultation with the PA manager. Decisions on activities to be undertaken by the eco-development committees are either taken unanimously in the general assembly where all the members participate or by the executive committee. The amount received can be used for the improvement of facilities in the villages as decided by the villagers.

Revenue is collected in the Vikas Nidhi since 1995–1996, but the PA began the process of revenue sharing with the eco-development committees in 2000–2001. From 2000–2001 to 2009–2010, the revenue generated was US$ 3,513,857, of which US$ 434,669 (12.37%) was provided to the eco-development committees toward a support fund.

The balance of the revenue is utilized for tourism infrastructure development, park management, and providing incentives to the staff. Since 2003–2004, each Forest Guard of the park has been getting US$ 97.99 per year as an incentive from the Kanha Vikas Nidhi. This is in addition to the salary received per month. Each daily wager, working in the park, besides his daily wages, receives US$ 76.21 per year as an incentive for their hard work in wildlife protection.

9.6.3 DIRECT CONTRIBUTION TO RURAL LIVELIHOOD

Authors have discussed the conservation benefits from nature tourism to PAs and local people (Lindberg et al., 1996; Kiss, 2004; West & Carrier, 2004; Nash, 2009; Sims, 2010; Sinha et al., 2012; Rastogi et al., 2015). Some suggest that nature-based tourism improves local livelihoods, park management, and promotes conservation by reducing pressures on forest resources (Goodwin, 1996; King & Stewart, 1996; Wilkie & Carpenter, 2002; Lindsey et al., 2006). Employment opportunities can be an incentive for community participation and cooperation in PA management (Holmes, 2003; Liu et al., 2012; Snyman, 2012b). Some communities close to parks have benefitted (Bookbinder et al., 1998; Nagendra et al., 2005; Spiteri & Nepal, 2008; Adams et al., 2010; Sinha et al., 2012).

It is widely accepted that biodiversity loss and poverty are linked problems and that conservation and poverty reduction should be tackled together (Adams et al., 2004). Tourism can provide a very suitable response to the need for economic regeneration and job creation in remote areas where traditional industries can no longer satisfy this need on their own (Azmi, 2005). The Guidelines for Ecotourism also emphasize the need to ensure benefits and income to local community beyond the park boundaries. In Periyar Tiger Reserve involving fringe communities in tourism has showed positive results. The increased income generated from resulted in reduction of poverty and reduced impact on park resources. Ecological monitoring of Periyar Tiger Reserve in 2000 indicated that regeneration of vayana (*Cinnamomum* species) has improved from about 6% to more than 13% and animal sightings in the tourism zone have increased (Uniyal & Zacharias, 2001).

The increase in visitation to the reserves has also resulted in an increase of tourism enterprises beyond the boundary of the reserve. Most of the economic benefits from these enterprises are limited to the villages located at the entrance of the park where tourism related facilities are clustered (Karanth & DeFries, 2011; Sinha et al., 2012; Bennett et al., 2012). As per NTCA Guidelines, all visitor related infrastructure have to be outside the reserve. Tourism infrastructure outside the Tiger Reserves is all in the hands of private entities. With the increase in tiger tourism the number of hotels and resorts solely catering to the park visitors have gone up from homestays to high-end premium hotels providing wildlife experience. In 1973–1974, there were 50 beds available in and around Kanha and in 2010 this rose to 1208 beds per night. Similar increases have been noticed around most of the Tiger Reserve. The tariff per night ranges from US$ 46 to US$ 1200 around Ranthambore, from US$ 15 to US$ 230 around Corbett, and from US$ 24 to US$ 492 in Kanha Tiger Reserve. The businesses of these hotels are for eight months a year, during which the park is open, and are directly dependent on visitors to the Tiger Reserve. But are the benefits reaped by the private businesses shared with the park and the people living around the reserve (Table 9.3)?

TABLE 9.3 Ownership of Resorts around Kanha Tiger Reserve.

Type of resort	No. of resorts	Resorts owned by locals	Resorts owned by outsiders
Low priced	13	4 (30)	9 (69)
Medium priced	8	1 (13)	7 (88)
High priced	8	0	8 (100)
Expensive	3	0	3 (100)

Figures in parentheses represent percentages.

The reserve provides employment opportunities for local communities in the neighborhood as watchers, guides, safari vehicle drivers in addition to working in the resorts and hotels, or own small shops selling daily use items, souvenirs, and eateries (Table 9.4).

TABLE 9.4 Employment Generated by Tourism for Community Living around Kanha.

Type	Number employed	Percentage of total population of seven villages surveyed
Resorts	292	7.05
Drivers/owners of vehicles	250	6.03
Guides	92	2.22
Shops	70	1.69

In Kanha, the monthly income of those employed in activities related to tourism is ~US$ 73 (Sinha et al., 2012).

The employment potential of tourism is low, and the direct impact of tourism on the household income is marginal. Only 19% of the population of the seven villages surveyed, which is 0.71% of the total population of the 150 villages around Kanha is engaged in tourism related activities. Thus it is clear that villages close to the hub of tourism, that is, the entry gates are more active in tourism (Table 9.5).

To define the economic status in society, the Government of India has pegged the poverty line at US$ 13.06 per month for rural areas (Government of India, 2009). It is estimated that ~750 million people continue to live in India's 600,000 villages, many of which are geographically and socio-culturally remote from major economic centers (Government of India, 2001). Communities living in the remote peripheral villages around PAs are poor, suffering livelihood insecurities and are heavily dependent on the forests for fuel, and fodder. Due to remoteness of the PAs there are no other source of livelihood for the communities apart from tourism in the reserves.

TABLE 9.5 Household Income from Tourism in Kanha Villages.

Village	Number of households surveyed	Households involved	Mean monthly income (US$)
Khatia	92	24 (26)	70
Sautia	35	04 (11)	92
Chapri	70	09 (13)	63
Mocha	99	27 (27)	70
Mukki	86	17 (20)	87
Manegaon	24	0	0
Patpara	13	0	0

Figures in parentheses represent percentages.

Households earning more than the others are those of vehicle owners who provide vehicles on hire to visitors for safaris. By rural standards, these households are affluent and belong to non-tribals. On average the safari vehicles charge US$ 32.66 per trip.

9.7 EDUCATION TO VISITORS: A NON-FINANCIAL CONTRIBUTION

Areas with natural beauty and ecological significance are increasingly concerned about promoting understanding and sensitivity to environmental issues (Moscardo & Pearce, 1986). Visitors often come with their own agendas of recreation and are non-captive audience from diverse backgrounds but as part of their wildlife tourism experience can be educated to increase conservation awareness and behave in ways that have positive consequences. As part of the tourism experience PAs incorporate nature interpretation and education through attended services like guides and unattended services like signages, publications, interpretation center, and visitor center. There is some evidence that interpretation associated with wildlife viewing can result in more positive attitudes toward wildlife conservation, in terms of generating increased support for conservation of target species and wildlife in general (Moscardo et al., 2001; Sinha et al., 2012).

Tourism in tiger reserves in India is able to achieve the objective of creating awareness amongst the visitors by interpreting the resources at several levels: through guided vehicle safaris, tiger shows on elephant back, and through interpretation centers. A study conducted, in two tiger reserves, to find out the effectiveness of nature interpretation in enhancing knowledge amongst visitors showed that the interpretation facilities helped visitors in gaining knowledge (Sinha et al., 2012). This was evident from the responses of visitors pre and post visit to the interpretation centers. Most of the visitors to the reserve are from urban background and bring with them prior knowledge and interest, are aware of conservation issues through other sources like television, films, documentaries, books, and so forth, before they come on the visit. Interpretation center helps in reinforcing their knowledge. This is evident as there is no significant difference in the answers to questions about general conservation issues before they visited the interpretation center and after. But there was a significant difference in the correct answers in pre and post visit for specific park related questions, which can be attributed to the exhibit displays in the interpretation centers.

The worth of an interpretation center is to be measured by what visitors carry back from it. In both the reserves, visitors responded that they learnt something from the interpretation center. The populace in India was more conservation oriented culturally, but voices against bad development have become stronger, to which visits to PAs have contributed significantly.

9.8 CONCLUSION

The overview presented a glimpse of the potential opportunity of tourism in Tiger Reserve in India and how it can be a viable economic activity if managed properly. But there are some threats too arising out of tourism which have impacts on tiger habitat and require regular monitoring. The way forward is an urgent need to manage specific problems arising out of tourism activity so that it can be beneficial to both the reserve and the community living around.

KEYWORDS

- **tiger reserve**
- **tourism**
- **contribution**
- **livelihood**

REFERENCES

Adams, V. M.; Pressey, R. L.;Naidoo, R. Opportunity Costs: Who Really Pays for Conservation? *Biol.Cons*. **2010,** *143,* 439–448.

Adams, W. M.; Infield, M. Who is on the Gorilla's Payroll? Claims on Tourist Revenue from a Ugandan National Park. *World Dev.* **2003,** *31,* 177–190.

Adams,W. M.; Aveling, R.; Brockington, D.; Dickson, B.; Elliot, J.; Hutton, J.; Wolmer,W. Biodiversity Conservation and Eradication of Poverty. *Science.* **2004,** *306,* 1146–1149.

Archabald, K.; Naughton-Treves, L. Tourism Revenue Sharing Around National Parks in Western Uganda: Early Efforts to Identify and Reward Local Communities. *Environ. Conserv.* **2001,** *28,* 135–149.

Azmi, N. TheEconomics of Tourism: Maximisingthe Benefits of Ecotourism for the Locality. *Environ. Sci.* **2005,** *9,* 43–52.

Beschta, R. L.; Ripple, W. J. Large Predators and Trophic Cascades in Terrestrial Ecosystems of the Western United States. *Biol. Cons.* **2009,** *142* (11), 2401–2414.

Billgren, C.; Holmén, H. Approaching Reality: Comparing Stakeholder Analysis and Cultural Theory in the Context of Natural Resource Management. *Land use policy.* **2008,** *25* (4), 550–562.
Bloom, P. Language and Thought: Does Grammar Makes us Smart? *Curr. Biol.* **2000,** *10,* R516–R517.
Bookbinder, M. P.; Dinerstein, E.; Rijhal, A.; Cauley, H.; Rajouria, A. Ecotourism's Support of Biodiversity Conservation. *Conserv. Biol.* **1998,** *12 (*6), 1399–1404.
Buckley, R. C.; Castley, J. G.; Pegas, F.; Mossaz, A. C.; Steven, R. A Population Accounting Approach to Assess Tourism Contributions to Conservation of IUCN-Redlisted Mammals. *PLoS ONE.* **2012,** 7, e44134.
Bushell, R.; McCool, S. F.Tourism as a Tool for Conservation and Support of Protected Areas: Setting the Agenda. In*Tourism and Protected Areas: Benefits Beyond Boundaries;* Bushell, R., Eagles, P. F. J., Eds.; CABI International: Wallingford, US, 2007.
Cleary, T. J. TheDevelopment and Validation of the Self-Regulation Strategy Inventory–Self-Report. *J. School Psychol.***2006,** *44,* 307–322. DOI: 10.1016/j.jsp.2006.05.002.
Clough, P. The Value of Ecosystem Services for Recreation. In *Ecosystem Services in New Zealand – Conditions and Trends;* Dymond, J. R., Ed.; Manaaki Whenua Press: Lincoln, New Zealand, 2013.
Cook, S. D.; Stewart, E.; Repass, K. *Tourism and the Environment;* Travel Industry Association of America: Washington DC, 1992; p 79.
Eagles, P. F. J. Global Trends Affecting Tourism in Protected Areas. In *Tourism and Protected Areas. Benefits beyond Boundaries;* Bushell, R., Eagles, P., Eds.; CAB International: Wallingford, UK, 2007; pp 27–43.
Eagles, P. F. J.; McCool, S. F.; Haynes, C. D. A. *Sustainable Tourism in Protected Area: Guidelines for Planning and Management;* IUCN: Gland, UK, 2002; xv + p 183.
Eagles, P. F. J. The Travel Motivations of Canadian Ecotourists. *J. Travel Res.* **1992,** *31* (2), 3–7.
Garnett, S. T.; Crowley, G. M.; Stattersfield, A. J. Changes in the Conservation Status of Australian Birds Resulting from Differences in Taxonomy, Knowledge and the Definitions of Threat. *Biol. Cons.* **2003,** *113,* 269–276.
Goodwin, H. In Pursuit of Ecotourism. *Biodivers.Conserv.* **1996,** *5,* 277–291.
Government of India 2001; Census of India. 2001.
Government of India 2009; Report of the Expert Group to Review the Methodology for Estimation of Poverty. Report by the Planning Commission, Government of India: India, 2009.
Hannam, K. Tourism Management Issues in India's National Parks: An Analysis of the Rajiv Gandhi (Nagarhole) National Park. *Curr. Issues.Tour.* **2005,** *8,* 165–180.
Holmes, C. M. Assessing the Perceived Utility of Wood Resources in a Protected Area of Western Tanzania. *Biol. Cons.* **2003,** *111,* 179–89.
Jamal, T.; Stronza, A.; Collaboration Theory and Tourism Practice in Protected Areas: Stakeholders, Structuring and Sustainability. *J. Sustain.Tour.* **2009,** *17,* 169–189.
Jhala, Y. V., Qureshi, Q., Gopal, R., Eds.; *The Status of Tigers in India 2014;* National Tiger Conservation Authority, New Delhi & The Wildlife Institute of India: Dehradun, India, 2014.
Karanth, K.K.; DeFries, R. "Nature-based Tourism in Indian Protected Areas: New Challenges for Park Management", *Conserv. Lett.* **2011,** *4,* 137–149.

Karanth, K. K.; Nepal, S. Local Perceptions of Benefits and Losses of Living around Protected Areas in India and Nepal. *Environ. Manage.* **2012,** *49,* 372–386. doi: 10.1007/s00267-011-9778-1.

Karanth, K. K.; Defries, R.; Srivathsa, A.; Sankaraman, V. Wildlife Tourists in India's Emerging Economy: Potential for a Conservation Constituency? *Oryx.* **2012,** *46,* 382–390.

King, D. A.; Stewart, W. P. Ecotourism and Commodification: Protecting People and Places. *Biodivers. Conserv.* **1996,** *5,* 293–305.

Kiss, A. Its Community-Cased Ecotourism a Good Use of Biodiversity Conservation Funds? *Trends Ecol. Evol.* **2004,** *19,* 232–237.

Leader-Williams, N.; Dublin, H. T. Charismatic Megafauna as 'Flagship Species'. In *Priorities for the Conservation of Mammalian Diversity: Has the Panda had its Day?;* Entwistle, A., Dunstone, N., Eds.; Cambridge University Press: Cambridge, UK, 2000; pp 53–81.

Lerner, R. M.; Lerner, J. V.; Almerigi, J.; Theokas, C.; Phelps, E.; Gestsdottir, S. Positive Youth Development, Participation in Community Youth Development Programs, and Community Contributions of Fifth Grade Adolescents: Findings from the First Wave of the 4-H Study of Positive Youth Development. *J. Early Adolesc.* **2005,** *25* (1), 17–71.

Lindberg, K.; Enriquez, J.; Sproule, K. Ecotourism Questioned-Case Studies from Belize. *Ann. Tour. Res.***1996,** *23,* 543–562.

Lindsey, P. A.; Alexander R.; Frank L. G.; Mathieson, A.; Romanach, S. S. Potential of Trophy Hunting to Create Incentives for Wildlife Conservation in Africa where Alternative Wildlife-Based Land Uses may not be Viable. *Anim. Conserv.* **2006,** *9,* 283–291.

Liu, W.; Vogt, C. A.; Luo, J.; He,G.; Frank, K. A.; Liu. J. Drivers and Socioeconomic Impacts of Tourism Participation in Protected Areas. *PLoS One.* **2012,** *7,* e35420. http://dx.doi.org/10.1371/ journal.pone.0035420.

Lyngdoh, S.; Shrotriya, S.; Goyal, S. P.; Clements, H.; Hayward, M. W.; Habib, B. Prey Preferences of the Snow Leopard (PantheraUncia): Regional Diet Specificity Holds Global Significance for Conservation. *PLoSONE.* **2014b,** *9,* e88349.

McCool, S. F. Challenges and Opportunities at the Interface of Wildlife Viewing, Marketing and Management in the Twenty First Century. In *Wildlife and Society: The Science of Human Dimensions;* Manfredo, M. J., Vaske, J. J., Brown, P. J., Decker, D. J., Duke, E. A., Eds.; Island Press: Washington, DC, 2009; pp 262–274.

Morrison, R. I. G.; Mizrahi, D. S.; Ross, R. K.; Ottema, O. H.; de Pracontal, N.; Narine, A. Dramatic Declines of Semipalmated Sandpipers on their Major Wintering Areas in the Guianas, Northern South America. *Waterbirds.* **2012,** *35,* 120–134.

Moscardo, G.; Pearce, P. L. Visitor Centres and Environmental Interpretation: An Exploration of the Relationships Among Visitor Enjoyment, Understanding and Mindfulness. *J. Environ. Psychol.* **1986,** *6,* 89–108.

Moscardo, G.; Woods, B .; Greenwood, T. Understanding Visitor perspectives on Wildlife Tourism. Wildlife Tourism Research Report Series No. 2., Australia, 2001.

Nagendra, H.; Karna, B.; Karmacharya, M. Cutting across Space and Time: Examining Forest Co-Management in *Nepal. Ecol. Soc.* **2005,** *10,* 24. [online] URL: http:// www.ecologyandsociety.org/vol10/iss1/art24/

Nagothu, U. S. Local People's Attitudes towards Conservation and Wildlife Tourism around Sariska Tiger Reserve, India. *J. Environ. Manage.* **2003,** *69,* 339–347.

Narain, U.; Shreekant, G.; vant Veld, K. *Poverty and the Environment-Exploring the Relationship between Household Incomes, Private Assets and Natural Assets;* Centre for Development Economics, Delhi School of Economics: Delhi, India, 2005; (Working paper 134).

Nash, S. Ecotourism and other Invasions. *Bioscience.* **2009,** *59,* 106–110.

Okello, N.; Beevers, L.; Douven, W.; Leentvaar, J. In *Breaking Kenyan Barriers to Public Involvement in Environmental Impact Assessment,* IAIA08 Conference Proceeding, The Art and Science of Impact Assessment 28th Annual Conference of the International Association for Impact Assessment, May 4–10, 2008; Perth Convention Exhibition Centre: Perth, Australia, 2008.

Rastogi, A.; Hickey, G. M.; Badola, R.; Hussain, S. A. Wildlife-Tourism, Local Communities and Tiger Conservation: A Village-Level Study in Corbett Tiger Reserve, India. *Forest Policy Econ.* **2015,** *61,* 11–19. http://dx.doi.org/10.1016/j.forpol.2015.04.007.

Salzer, D.; Salafsky, N. Allocating Resources between Taking Action, Assessing Status, and Measuring Effectiveness of Conservation Actions. *Natural Areas.* **2006,** *26,* 310–316. http://dx.doi.org/10.3375/0885-8608(2006)26 [310:ARBTAA]2.0.CO;2.

Sandbrook, C. Putting Leakage in its Place: The Significance of Retained Tourism Revenue in the Local Context in Rural Uganda. *J. Int. Dev.* **2008,** *22* (1), 124–136.doi: 10.1002/jid.1507.

Sims, K. R. E. Conservation and Development: Evidence from Thai Protected Areas. *J. Environ. Econ. Manage.* **2010,** *60,*94–114.

Snyman, S. 'The Role of Tourism Employment in Poverty Reduction and Community Perceptions of Conservation and Tourism in Southern Africa.' *J. Sust. Tour.* **2012a,** *20* (3), 395–416.

Snyman, S. 'Ecotourism Joint Ventures between the Private Sector and Communities: An Updated Analysis of the Torra Conservancy and Damaraland Camp Partnership, Namibia.' *Tourism Manage.Perspectives.***2012b,** *4,*127–135.

Spiteri, A.; Nepal, S.K.'Evaluating Local Benefits from Conservation in Nepal's Annapurna Conservation Area.'*Environ. Manage.* **2008,** *42,*391–401.

Steven, R. Tourism: A Threat to Birds and a Tool for Conservation. Honours Thesis, Griffith School of Environment, Griffith University, Gold Coast, Australia, 2011, p 98.

Tiger Task Force Joining the Dots (TTF). Union Ministry of Environment and Forests (Project Tiger), 2005.

Tosun, C. Towards a Typology of Community Participation in the Tourism Development Process. *Int. J. Tour. Hospitality.***1999,** *10*, 113–134.

Tosun, C. Limits to Community Participation in the Tourism Development Process in Developing Countries. *Tour. Manage.***2000,** *21,* 613–633.

Sezgin, E.; Yolal, M. Golden Age of Mass Tourism: Its History and Development. In *Chapter from the Book Visions for Global Tourism Industry - Creating and Sustaining Competitive Strategies;* INTEC Open Science: Croatia, 2012.

Sinha, B. C.; Qamar, Q.; Uniyal, V. K.; Sen, S. Economics of Wildlife Tourism – Contribution to Livelihoods of Communities around Kanha Tiger Reserve, India. *J. Ecotour.* **2012,** *11* (3), 207–218. DOI:10.1080/14724049.2012.721785.

West, P.; Carrier, J. G. Ecotourism and Authenticity-Getting away from it all? *Curr. Anthrop.* **2004,** *45,* 483–498.

Wilkie, D. S.; Carpenter, J. F. Can Nature Tourism Help Finance Protected Areas in the Congo Basin? *Oryx.* **2002,** *33,* 333–339.
Winterbach, H. E. K.; Winterbach, C. W.; Somers, M. J.; Hayward, M. W. Key Factors and Related Principles in the Conservation of Large African Carnivores. *Mammal Rev.* **2013,** *43,* 89–110. DOI 10.1111/j.1365-2907.2011.00209.x.

CHAPTER 10

MANAGING WILDLIFE TOURISM IN SRI LANKA: OPPORTUNITIES AND CHALLENGES

JEREMY BUULTJENS[1*], IRAJ RATNAYAKE[2], and ATHULA CHAMMIKA GNANAPALA[2]

[1]*School of Business and Tourism, Southern Cross University, Lismore, New South Wales, Australia*

[2]*Department of Tourism Management, Sabaragamuwa University of Sri Lanka, Belihuloya, Sri Lanka*

[*]*Corresponding author. E-mail: Jeremy.buultjens@scu.edu.au*

CONTENTS

ABSTRACT

The Sri Lankan wildlife tourism sector, given the diverse range of wildlife, many of which are endemic, is well placed to attract high yield tourists. Currently, the majority of wildlife tourism experiences occur within the country's protected areas (PAs). The potential benefits arising from the expansion of this sector are likely to be significant however without proper management there could also be a number of deleterious impacts. The problem for the governing agency, the Department of Wildlife Conservation (DWLC), is to effectively manage the delicate balance between the obligations to protect environmental values whilst at the same time provide quality recreational opportunities. The rapid increase in international tourism since 2009 is likely to exacerbate this difficulty for the DWLC. It is estimated that ~30% of international tourists visit a national park (NP) during their stay in the country. In addition, the level of domestic tourism is expected to increase thus putting additional pressures on NPs. This chapter examines the opportunities and challenges associated with wildlife tourism. The data for this study is based on a review of the literature and documents relating to wildlife tourism in Sri Lanka in addition to semi-structured, personal interviews with nine travel and tour executives, three park managers, and one ex-DWLC officer. It is clear that increasing visitation numbers can contribute to the economic development of the country and regions where the parks are based as well as providing sources of the funding for maintaining environmental values of the parks. However, at the present time there are negative environmental impacts occurring as a result from tourism. These impacts are also reducing the amenity value for visitors. The DWLC is attempting to use various strategies to ameliorate the negative impacts caused by visitation however their efforts are hamstrung by the lack of funding.

> **".....postulated wildlife tourism may only be sustainable if there are benefits for the local communities."**

10.1 INTRODUCTION

Sri Lanka, with its diverse range of wildlife, including many endemic and charismatic species such as elephants, leopards, sloth bears, golden jackals, mongooses, and whales, is potentially well-placed to capture considerable benefits for itself and its wildlife (Lai, 2002; see also Higginbottom, 2004). From a tourism perspective, these diverse range of species provide the country with a "truly unique competitive advantage" (de Silva Wijeyeratne, 2006a). The island's wildlife assets have made it one of the 25 biodiversity hotspots in the world (Ministry of Environment and Natural Resources, 2002). Currently, the majority of non-captive wildlife tourism experiences occur within the country's protected areas (PAs) that include the 22 national parks (NPs) (including the two marine parks and two World Heritage Sites) (DWLC, n.d.). In 2012, terrestrial PAs covered 22.0% of the country's 65,610 Sq. km of the land area (World Bank, n.d.). The PAs are managed by Department of Wildlife Conservation (DWLC) (n.d.), and Forest Department (FD) (Senevirathna & Perera, 2013).

In addition to non-captive experiences, there are also captive wildlife tourism opportunities including the popular Pinnawala, a state-run elephant orphanage established in 1972, and the Zoological Gardens, in Colombo (Jayantha, n.d.). Private operators also provide wildlife tourism opportunities such as those offering turtle hatching experiences previously funded by Non-governmental organizations (NGOs) (Tisdell & Wilson, 2012) and elephant-back safari tours (Jayantha, n.d.). Finally, some wildlife tourism opportunities are provided by certain pageants or peraheras held throughout the country, such as the Kandy perahera which involves a parade of elaborately dressed elephants. Many of the captive forms of wildlife tourism have attracted substantial criticism (see de Silva Wijeyeratne, 2010; Jayantha, n.d; Tisdell & Wilson, 2005a) however they are outside the scope of this chapter. This chapter only considers the perceived potential of the non-captive component of the land-based[1] sector and some of the challenges and issues that may threaten its sustainability. These insights, provided by a number of stakeholders, will result in a better understanding of the sector and this should facilitate improved planning for its development.

10.2 WILDLIFE TOURISM

Wildlife tourism, that is, tourism based on encounters with non-domesticated (non-human) animals in either the natural environment or in captivity

[1]A description of the marine-based wildlife tourism is provided in Buultjens et al. (Under review).

(Higginbottom 2004) has close links with nature-based and ecotourism; they are often used interchangeably although there are important differences between them. For example, nature-based tourism activities, that is, activities occurring in a natural setting, may or may not include encounters with wildlife. Wildlife tourism experiences can include deliberate and accidental animal encounters in a natural setting as well as encounters in non-natural settings such as zoos, theme parks, and aquaria (Fredline, 2001). Finally, ecotourism, while incorporating aspects of both nature-based and wildlife tourism, also has some unique features such as a focus on education and equity (Sharpley, 2000).

Regardless of its setting and characteristics, wildlife tourism can provide short- and long-term benefits for wildlife and the environment as well as providing economic and social benefits for local communities. Wildlife tourism also enhances a destination's appeal and visitor experience (Ballantyne et al., 2011; Tisdell & Wilson, 2005b). Importantly, by providing socio-economic benefits, wildlife tourism, especially the non-captive form, provides an incentive for the ongoing protection of a country's wildlife and their habitats (Ballayntyne et al., 2011; Higginbottom et al., 2001; Wilson & Tisdell, 2001). The industry can also assist in developing respect and appreciation for wildlife and nature amongst hosts and guests as well as raising awareness of environmental issues (see Ballayntyne et al., 2011). The ongoing delivery of these benefits requires effective sustainable management that protects wildlife and habitat as well as meeting the needs of hosts and guests (Rodger et al., 2010). In the absence of sustainable management there is the potential for serious problems to arise (Banerjee, 2012), which, depending on the species, life-cycle stages and habitats, include stress caused by close contact with humans, injury or death of wildlife, habitat alteration, modification of natural behavior, overfeeding, and pollution. There are also risks of injury to humans (Green & Higginbottom, 2001; Newsome et al., 2005). In addition, in situations where the local community are not properly engaged in planning they are likely to experience negative impacts from the industry as well as an inequitable return (Higginbottom, 2004). The visitor experience of wildlife tourists can also be compromised without proper planning, for example, overcrowding at sites. Any increases in the number of wildlife tourists is likely to further exacerbate potential problems and the involvement of threatened species and/or travel to increasingly remote areas further aggravates the situation (STCRC, 2009).

In addition to impacts on wildlife from tourism there can also be impacts that arise from outside the industry. For example, many iconic species, including elephant and rhinoceros, are declining due to excessive illegal killing and trade (Collaborative Partnership on Sustainable Wildlife Management (CPW), 2014). In order to overcome these non-tourism impacts on wildlife (and therefore ultimately on tourism) there is a need to garner the support of host communities. This can sometimes be difficult since the tourism industry can also impact negatively on host communities. In order to gain local support it is important to ensure host communities benefit from tourism while any negative impacts on them are ameliorated. Not surprisingly the level of community support and attitudes toward conservation are determined by the level of benefits people receive from the industry and PAs (Udaya, 2003). Indeed, Burns and Schofield (2001) note that it is postulated wildlife tourism may only be sustainable if there are benefits for the local communities. Unfortunately the benefits from wildlife tourism are often not distributed equitably (Adams & Infield, 2003; Shackley, 1996).

It is not only important for the local community to benefit from tourism, managing agencies need adequate funds to protect environmental values and the tourism activities taking place within them. Unfortunately many managing agencies do not have the sufficient funding and/or capacity to provide effective management. Many of these agencies also often suffer from poor governance, corruption, and lack of political will (Buultjens et al., 2005; CWP, 2014). These factors can result in the design and adoption of inadequate and/or inappropriate policies that are aimed at supporting sustainable management (CWP, 2014). Poor management regimes are further exacerbated by a lack of a sound understanding of biological and ecological elements, such as species' habitats, population sizes, migration routes, and population demographics (CWP, 2014). A lack of a good knowledge is unfortunate since wildlife management decisions need to be informed by science. It is also important that all stakeholders are also cognizant of the science. Furthermore, decisions need to be made in a structured and transparent manner (British Columbia Ministry of Environment, n.d.).

In establishing appropriate strategies and actions to address and ameliorate negative wildlife tourism impacts there is a need to careful consider their likely effectiveness (Tisdell & Wilson, 2004). For example, the introduction of entry fees to NPs may draw criticism and possibly impact negatively on the tourism industry however this criticism can be muted if the revenue collected from park entry is used to improve amenities and conservation in the parks. In this situation the use of fees will be beneficial for PA management as well as for the tourism industry (see Tisdell & Wilson, 2004).

In addition to ameliorating environmental impacts, effective wildlife tourism management needs to generate funds for conservation within and outside of PAs (CWP, 2014) and the provision of benefits to local communities. Successful management will also ensure an effective education and interpretation program. It is recognized that successful biodiversity conservation and tourism management is dependent on education and public awareness amongst all stakeholders (Banerjee, 2012; CWP, 2014). Effective environmental management requires that all sectors of society including individuals, organizations, communities, industries, and government to acknowledge that they have a responsibility in management. A successful shared stewardship model involves information and knowledge sharing, consultation, sustained support, and the development of partnerships (British Columbia Ministry of Environment, n.d). The shared stewardship model recognizes "the importance of the human dimension, not only in terms of people's needs and benefit-sharing, but also with respect to generating incentives and funding for wildlife conservation and sustainable use" (CWP, 2014).

10.3 TOURISM IN SRI LANKA

The establishment of the Ceylon Tourist Board in 1966 and the release of the Tourism Management Plan in 1967 (Ceylon Tourist Board, 1968; Jayawardena, 2013) is often perceived as the beginning of the tourism industry. Between 1970 and 1980, annual growth in international visitation was in excess of 21% (Sri Lanka Tourism Development Authority, 2005). This growth was interrupted by the 25-year-old civil war where international visitation stagnated at between 400,000 and 500,000 visitors. This period was also characterized by some intermittent and sharp falls in visitation as a consequence of specific terrorist acts and the tsunami that struck in 2004 (Sri Lanka Tourism, n.d.). Since cessation of the war international visitation has grown rapidly increasing from 447,890 tourists in 2009 to over 1.5 million in 2014 (Sri Lanka Tourism, n.d.).

Until relatively recently, the tourism industry has been almost exclusively dependent upon the comparatively low-yield, mass international "sea, sun and sand" market (Buultjens et al., 2015; Fernando et al., 2013). This is not surprising given its tropical weather and substantial coastline. The other traditionally popular tourism product has been the cultural experiences offered in the so-called "cultural triangle" based around Kandy, Anuradhapura, and Polonnaruwa. This area contains five of the country's seven world heritage sites (Ranasinghe, 2015). Despite the popularity of these two segments, since 2000, the government and the industry have acknowledged the need for industry to diversify from its historically narrow focus to include more high-yield markets, including ecotourism, nature-based and wildlife tourism (Buultjens, 2014; Perera & Vlosky, 2013). An example of an initiative to diversify was declaration of the year 2000 as the "Year of Ecotourism" by the Ministry of Tourism (Fernando & Shariff, 2013). Another example was the launch of the *Refreshingly Sri Lanka–Visit 2011* marketing campaign by Sri Lanka Tourism that highlighted 12 experiences including heritage, nature, and wildlife. The drive for diversification of the industry received further impetus from the rapid expansion of international visitation since the end of the Civil War in 2009 and the realization that there was a need for a greater range of tourism product (Anon, 2011; Buultjens et al., 2015; ICRA Lanka, 2011).

Despite this drive for diversification and the attractiveness of country's wildlife, this segment (eco and nature-based tourism) is still in its infancy with only a relatively small percentage of international tourists actually visiting a NP during their stay. However, the number is growing relatively quickly. For example, in 2011, only 23% (198,536) of the 855,975 international tourists visited a NP (Miththapala, 2012). By 2013, this figure had grown to 325,153 representing just fewer than 30% of all visitors. The rapid increase in international visitation since the end of the war as well as the increased promotion of and interest in wildlife, nature-based and ecotourism is likely to result in considerable extra pressure on the country's PAs.

10.4 WILDLIFE AND PROTECTED AREAS IN SRI LANKA

As noted earlier, Sri Lanka has considerable biodiversity assets of which many are important tourism drawcards (de Silva Wijeyeratne, 2006a; Senevirathna & Perera, 2013). However, major concerns are being raised about the level of biodiversity loss (Sathurusinghe, 2008) with many of the many endemic fauna and flora under threat of extinction (Geekiyanage et al., 2015). This loss is the result of various factors including forest degradation, the collection of wild species for commercial purposes, introduction of invasive species, draining of wetlands for development purposes, destruction of mangrove forests and coral reefs, and hunting and poaching (Sathurusinghe, 2008). Biodiversity loss continues despite the existence of a relatively extensive system of PAs.

The system of PAs started when the first reserves—Yala (160 sq. miles) and Wilpattu (256 sq. miles)—were established in 1900 and then later designated as NPs in 1938 (DWLC, n. d.). These initial reserves have grown into an extensive system covering over 22% of the country's total land area. The majority of these areas (87%) are administered by the DWLC while the remaining are administered by the FD; both of which are under the responsibility of the Ministry of Environment and Natural Resources. The FD manages about 18% of the total land area for various purposes with over 45% of the existing natural forests allocated as PAs, specifically for protection and conservation purposes (Ministry of Economic Development, 2011; Sathurusinghe, 2008). The PAs under the FD can be categorized as either (a) World Heritage sites—for example, Sinharaja; (b) National Heritage and Wilderness Areas—for example, Knuckles; (c) Man and Biosphere Reserves—three International Reserves and 47 National Reserves; and (d) Other Conservation Forests (Sathurusinghe, 2008). The FD management allow research, education, and recreation to be undertaken in their PAs and these areas appear to be highly compatible with nature-based/wildlife tourism (Senevirathna & Perera; 2013).

The DWLC has responsibility for 22 NPs that cover over 526,000 hectares. In addition there are three Strict Nature Reserves, four Nature Reserves, 61 Sanctuaries and one jungle corridor covering an additional 446,513 hectares (Senevirathna & Perera, 2013). Strict Nature Reserves are highly protected landscapes where only research and educational activities are allowed with the permission of the Director General of the DWLC. Nature Reserves are also highly protected however traditional human activities are allowed within their boundaries. In contrast NPs, while protecting wildlife and their habitats, provide opportunities for the public to observe and study wildlife. The DWLC (n.d.) note that one of its main functions is to facilitate ecotourism in PAs. Senevirathna and Perera

(2013) suggest that NPs are highly compatible with both hard and soft ecotourism activities. Finally, sanctuaries can be declared on state and private lands and require no permission or entry fee. These areas are also compatible with nature-based/wildlife tourism; however, wildlife viewing opportunities are comparatively limited due to the nature of their establishment (Senevirathna & Perera, 2013).

10.5 METHODS

This study utilized a mixed methods research design (Mertens, 2005) to obtain a comprehensive understanding of wildlife tourism in Sri Lanka. Firstly document analysis was undertaken to gain an overview of the sector and included examination of government policy documents, government media releases, academic journals and books, and commentary from a range of websites.

In addition to the document analysis, a series of in-depth interviews were conducted with various relevant stakeholders over a period of three-month period between July and September, 2015. Participants for this study were purposefully selected (Patton, 1990) with researchers aiming to select respondents who were likely to represent the views of stakeholders, that is, people who had direct dealings with wildlife tourists and/or with managing their behavior. Face-to-face, semi-structured interviews were considered the most appropriate method since they allow the time and opportunity for an in-depth exploration of ideas and the collection of rich data (Gillham, 2000). By the end of the interview period 13 stakeholders had been interviewed since there had been "saturation" of data collected (Charmaz, 2006). There were nine travel and tour executives interviewed, three park managers and one ex-DWLC officer who worked in the Tourism Division of the organization.

The in-depth interviews were conducted to determine the views of key stakeholders regarding the current position of the wildlife tourism sector, its potential and the issues it is facing. These interviews usually occurred at a place of convenience for the interviewee which in most cases was their workplace. No monetary benefits were provided to the study participants.

Prompting questions were asked to develop discussion and ensure the interviews covered the range of subject areas intended; however for the most part, the interviewers allowed participants to bring up concerns or topics without prompting to examine which themes naturally arose. Each interview lasted approximately one hour. The interviews were based around the following themes: the potential of wildlife tourism; measures to effectively promote the sector; the level of knowledge about sustainability of PAs amongst wildlife tourists; the perceived problems of wildlife tourism; the use of park fees and other management strategies aimed at facilitating the development of wildlife tourism; the involvement of local communities in wildlife tourism; delivering of benefit to the local community through wildlife tourism and the extent of education and interpretation.

10.6 WILDLIFE TOURISM IN SRI LANKA

10.6.1 VISITOR NUMBERS

It is not possible to identify which visitors are "wildlife" tourists; however, it is reasonable to assume that most people visiting a NP are doing so to view wildlife. Traditionally PAs have been relatively underutilized by the tourism industry however since the end of the war in 2009 they have started attracting increasingly more international and domestic tourists (Rathnayake & Gunawardena, 2011; Senevirathna & Perera, 2013). In 2009, 434,802 people visited the country's terrestrial NPs with international visitors accounting for only 16.3% of all visitors. The total revenue from visitation to terrestrial NPs amounted to Rs. 119,419,746.8 (US$ 845,884) with international visitors contributing 87.1% of revenue (Sri Lanka Tourism Development Authority, 2009). By 2014, 1,044,222 people visited the country's NPs, with international visitors accounting for 31.1% of all visitors. The total revenue from visitation has risen to Rs. 614,283,535.07 (US$ 4,351,144) with international visitors contributing 94.2% of revenue (Sri Lanka Tourism Development Authority, 2014). The figures indicate that there has been over a 140% increase in visitation and 414.4% increase in revenue.

All participants agreed that the wildlife sector is underdeveloped in Sri Lanka and they also saw a distinction between "pure" wildlife tourists and "other" tourists who have a wildlife experience. For example, the Tour executive for ecotourism nine (personal communication) noted "[T]he watching of wild animals through safaris cannot be called as wildlife tourism - I hope the definition of wildlife tourism is more than that [sick]." Pure wildlife tourists represent only a very small sector of the tourism market. The owner/tour executive 6 (personal communication) estimated that between 3 and 5% of tourists come for pure wildlife tourism. This is similar to the 5% suggested by Fernandopulle (2011). An airport survey in 2013 of departing tourists identified that 2.5% visited the country for the prime purpose of viewing wildlife. Another 6.2% identified wildlife as the second purpose and 11.5% respondents ranked wildlife the third purpose of their visit (Sri Lanka Tourism Development Authority, 2013). Pure wildlife tourists are perceived to be high spending visitors—they visit for longer and like to stay in luxury accommodation including luxury tents. They are also perceived as having a lower impact on the environment as well as having higher expectations about the facilities and experiences on offer.

In addition to the "pure" wildlife tourists there are many other visitors who seek wildlife, nature and culture experiences (Managing Director/Owner Travel Agency 2, personal communication). There was a consensus amongst participants that nearly all the travel agencies offer attractive nature/wildlife tourism experiences however very few of them offer pure wildlife tourist products. Wildlife tourism is usually offered as part of tour package; specifically safaris in NPs. The feeling amongst interview participants is reflected in the comment that "there are very popular and competing wildlife destinations globally, therefore (the) word wildlife comes to mind and (people) remember other countries other than Sri Lanka, therefore if we try to promote pure wildlife only it will not be an easy task, therefore most of the travel agents tend to promote mixed product/packages [sic]" (Tour executive for ecotourism 9, personal communication).

The interview participants identified Germans, Belgiums, French, and the British as the nationalities most interested in wildlife tourism. "I handle the UK market and 99% of the tourists like wildlife experiences therefore we include at least two wildlife safaris for the itinerary. In addition to the wildlife safaris in NPs we do promote trekking in some NPs, forests and other trekking areas developed by private businessmen to see butterflies, birds and other small animals etc. …Some hoteliers have developed their own tracks to give wildlife and nature experiences to their in-house guests with the support of a naturalist (who is normally provided by resort hotels)" (Travel executive 7, personal communication).

In contrast, tourists from the Middle East, India, Pakistan, and China are perceived as being much less interested in undertaking a wildlife experiences. The Travel executive for the Middle East/Pakistan market

1 (personal communication) suggested that this is "due to their social and cultural background – they do not like animals/insects even when they see some insects even in hotels they may complain [sic]." The Tour executive for the Indian market 4 (personal communication) noted that while Indian tourists are not that interested in wildlife they often visit Horton Plains NP. The NP is associated with the legendary story of Rama-Sitha and as a part of the Ramayana trail Indian tourists like to visit the park. The Owner/Tour executive 5 (personal communication) also noted that Russian tourists in general do not like visiting NPs and that they prefer to spend their time at the beach.

A number of participants noted that whale watching is becoming an increasingly popular activity amongst tourists including those from the Middle East. As with terrestrial wildlife tourism, whale watching is generally undertaken as part of an organized tour of the country. Another important component of wildlife tourism is bird watching. "We have separate packages for only focusing bird watching in Thalangama (private land), Sinharaja, Bundala, and Kumana. We add Kumana for tourists who are coming to stay normally stay more than 15 nights [sic]" (Owner/Tour executive 5, personal communication).

The interview participants observed that some parks are heavily used for tourism. For example, participants acknowledged Yala as the best known and most heavily visited park while Udawalawe, Minneriya, and Wilpattu NPs are also popular.

All participants suggested that while currently wildlife tourism is a relatively small sector demand is growing. There was considerable enthusiasm for the sector amongst the participants however most felt that Sri Lanka has not been well-positioned as a wildlife tourism destination by the government and the industry. "When it comes to wildlife tourism, tourists get 'Africa' into their minds. So it is difficult to attract pure wildlife tourists as Sri Lanka is famous for 'compact' – having everything [sic]" (Tour executive 3, personal communication). There was a feeling amongst interviewees that wildlife tourism is not promoted since there is a lack of proper infrastructure and qualified staff servicing the sector (these issues are examined later in the chapter). Another perceived issue is the difficulty in assuring visitors a sighting of many of the wildlife drawcards.

Travel executive 7 (personal communication) also noted that some tour operators were discouraging visitation to captive wildlife attractions such as domesticated animal orphanages and that increasingly more tourists, especially those from Europe, "are talking about animal cruelty and their rights, therefore tourists like to see the animals in natural setting."

10.6.2 VISITOR MANAGEMENT AND FEES

Visitation in all terrestrial NPs is allowed between 6.30 am and 6.30 pm daily. All visitors, other than those to Horton Plains, are required to hire a jeep or boat in the case of marine environments. In addition, all motor vehicles must be accompanied by a DWLC guide who often only speaks rudimentary English. Visitors are not allowed to alight from their vehicle. Currently whale watching boats are not accompanied by a DWLC guide although there are plans for this in the future (Buultjens et al., 2016).

The entry fees charged vary between the parks and between international and domestic visitors. International visitors are charged ~US$ 10/person at the less popular parks, up to US$ 15 at Yala and Udawalawe and US$ 20 at Horton Plains. Domestic visitors are charged around US$ 0.25. There are also additional charges including a "service charge" ($ 8/vehicle), which covers the services of the tracker, a "vehicle charge" (US$ 1.78)/vehicle); plus a 15% tax on all services (the exact entrance cost/person thus becomes slightly cheaper the more people sharing a vehicle). The charges for the jeep hire and its driver from a tour company range from ~US$ 30–45 for half a day hire to US$ 60–80 for a full day (Roughguides, n.d.).

Visitors also have the opportunity of staying in simple bungalows in many of the terrestrial NPs. The basic bungalow fee is ~US$ 25–30/person. In addition visitors have to pay a two-day park entrance fee, other add-on fees for a "service charge" and "linen charge," and an all-encompassing 15% fee. It typically costs around US$ 150/night for two people to stay in a park bungalow excluding transport costs to, from and around the park. Camping is also allowed in NPs and costs include a two-day entrance fee, a US$ 15 in camping fee as well as transport costs. Additionally, some tour companies also operate camping trips to various NPs (Roughguides, n.d.).

10.6.3 REVENUE AND FUNDING

As noted earlier the total revenue accruing to DWLC from visitation is substantial, especially from international tourists who contribute over

94% of revenue (Sri Lanka Tourism Development Authority, 2014). Despite this revenue, Rathnayake and Gunawardena (2011) assert that park entrance fees do not reflect the true value of the resource as a consequence of market failures and imperfections. The undervaluing of the resource is unfortunate since it results in its mismanagement (Rathnayake & Gunewardena, 2011). A lack of funding and increasing visitation to NPs has resulted in significant impacts making it more difficult for the DWLC to fulfill its dual mission of providing high quality recreational experience and conserving environmental resources for future generations.

Another factor that impacts on the provision of funding is the concentration of visitation in some of the better-known parks (Senevirathna & Perera, 2013). For example, in 2009, Yala NP attracted 27.5% of all visitors and by 2014 this had grown to 36.3%. The top four NPs in 2009 accounted for 88.2% of total visitation; by 2014, while visitation had dispersed slightly, the top four NPs still accounted for 75.9% of total visitation. This concentration means there is substantial difference between revenues generated for management at the different NPs. The less popular parks will have even fewer funds for park maintenance and conservation efforts (Senevirathna & Perera, 2013). Parks fees was not raised as an issue by interview participants suggesting that there is probably room for them to be increased. This would raise more funds for park management as well as possibly mitigating overcrowding.

10.6.4 WILDLIFE TOURISM ISSUES

The literature and interviews identified a number of issues that are impacting on the wildlife tourism industry in Sri Lanka.

10.6.5 WILDLIFE TOURISM VEHICLES

All participants noted that there were substantial problems with the safari jeeps used for transporting tourists to and through NPs. "We receive complaints from tourists regarding the quality of Jeep especially the comfort ability – very old jeeps are used" (Travel executive for the Middle East/Pakistan Market 1, personal communication). These older jeeps are very noisy and often scare wildlife reducing the probability of their viewing by tourists. The vehicles are also very polluting. It was noted that newer jeeps are increasingly being introduced but these raise the costs for tourists.

10.6.6 STAFFING INCLUDING DRIVERS AND GUIDES

In addition to the quality of the jeeps used for wildlife tourism the behavior of the drivers and guides received unanimous criticism from the interviewees. It was noted that tourists often criticize drivers for their poor dress and appearance as well as for their careless driving and speeding in the parks. There was another issue that concerned many interview participants. There is a practice within most parks that when an animal is spotted the driver calls the owner of the jeep (who then communicates the location to the company's other drivers) or directly communicates with the other drivers. This results in a number of jeeps rushing at high speed to the where the spotting took place. This practice results in high congestion, poor viewing of wildlife, and reduced visitor satisfaction. There can also be negative impacts for the wildlife including some being killed. For example, a leopard was killed by vehicles in Yala in August, 2015. In an attempt to prevent this from happening the DWLC recently banned jeep drivers bringing in mobile phones to Yala NP during a safari (Managing Director/Owner Travel Agency 2, personal communication). It is also planned that every safari jeep will be provided with global positioning system (GPS) trackers and closed circuit television (CCTV) cameras will be installed at selected locations throughout the parks (Park manager 12, personal communication).

There was also considerable criticism of the wildlife guides and this was seen as a major barrier to the promotion of wildlife tourism, especially camping in NPs. There can be a high turnover of guides since they may find employment elsewhere (Buultjens et al., 2005). The lack of qualified and trained guides also meant there is poor interpretation of the wildlife. The Owner/Tour executive 6 (personal communication) noted that the "majority of the workers engaged in wildlife (tourism) don't have knowledge regarding the concept of tourism and they don't have any professional skills to cater to the tourists' requirements. Furthermore, they don't even get a proper training." The Travel executive 7 (personal communication) suggested that it was difficult to find qualified trekkers in NPs and while most can provide advice on directions, they cannot guide or provide interpretation.

A number of service staff used by tour companies is drawn from local villages. These people are usually poorly educated and lack the proper skills, and have little access to proper training.

10.6.7 OVERCROWDING

Buultjens et al. (2005) found that in 2001 Yala NP was experiencing substantial overcrowding and that the park's carrying capacity had probably been exceeded. This assessment was based on the 153,661 wildlife tourists who visited in 2000. In addition to the officially recorded tourist numbers there were a further estimated 400,000 religious pilgrims who visited Yala in that year. In 2013 the number of visitors had grown to 379,414 tourists who visited the park; this represents a 147% increase in numbers since 2000. During the same period it would be safe to assume that the number of pilgrims would have either remained the same or increased. The increase in visitation pose a serious concern for park management and while Yala is the most popular NP in the country it is likely a number of other parks would be feeling the impact of the increase in tourism since the ending of the war.

The majority of interview participants supported the idea that overcrowding, especially in the more popular parks, was a major problem. Overcrowding was also seen as substantially impacting on the visitor experience. The impacts from overcrowding were exacerbated by the practices of the tour drivers outlined above where animal spotting is communicated amongst the drivers. The DWLC are working with jeep-owners associations to restrict the numbers of jeeps entering parks (Park Manager 12, personal communication).

10.6.8 FACILITIES

The lack of appropriate facilities within NPs was also identified as a major issue by most participants. Roads, toilets, park bungalows, and camping sites do not meet the expectations of tourists. Tour executive 3 (personal communication) noted that camping clients are high-end tourists who spend more than the average tourist. They also have high expectations and "compare their experiences with what they have paid. Here, always there is a gap in between these two [sic]." Another tour executive (Travel executive 8, personal communication) noted their company does not allow its clients to stay in the circuit bungalows in NPs due to lack of facilities.

The Tour executive for the Indian market 4 (personal communication) also noted that it is difficult to find appropriate accommodation in some towns positioned close to NPs. This becomes more of a problem when the park is more remote. As a result operators tend to use NPs that are closer to tourist hubs; further contributing to overcrowding in some parks.

10.6.9 DWLC

Buultjens et al. (2005) identified the DWLC's lack of resources as an important problem in 2001 and this continues to be a problem. The lack of human and financial resources is seen to result in the provision sub-par facilities and inefficient management of tour operators, guides, and visitors. The Tour executive for the Indian market 4 (personal communication) suggested that the existing policies aimed at protecting wildlife were satisfactory but that the organization did not have the resources to implement them effectively. It was felt that the DWLC needed to provide greater supervision especially to control the behavior of safari jeep drivers and trekkers. Owner/Tour executive 6 (personal communication) noted that there was no proper monitoring systems in place to evaluate the impacts of operations within the parks, although this may change in the future. One participant suggested that many of the malpractices occurring within the wildlife parks was due to illegal money and other benefits supplied to the authorities in order to get them to change their policies.

Another perceived problem with the DWLC was that some of the participants felt that, despite having an aim of facilitating tourism within NPs, the department was not always supportive of the promotion of wildlife tourism. The DWLC was seen as trying to use existing rules and regulations to limit the expansion of the industry. As one participant noted "[W]e are having many natural resources however we do not have capable human resources to utilize wisely and sustainably to get the expected benefits [sic]" (Tour executive 3, personal communication).

The DWLC participants were well aware of the problems involved in managing tourism in NPs. In one response to some of the problems the DWLC introduced a range of controls for visitors to Yala National Park in September, 2015. These controls will be legally enforced and include: a maximum speed of 25 km for all vehicles; a tour guide accompanying all visitor vehicles; the prohibition of guns and knives; discouragement of visitors bringing matches, lighters, cigarettes, and alcohol; a prohibition on feeding animals in the park; and prohibition of throwing garbage thrown from vehicles.

10.6.10 INTERPRETATION AND EDUCATION

The majority of participants believed that the interpretation and education provided to wildlife tourists was of a poor standard and did not meet the needs or the expectations of tourists, especially the "pure" wildlife tourists. As noted previously staff lack training and expertise as well as the language skills to provide effective interpretation. Sinhala and English are often used in conjunction with Sinhala and English publications despite the fact that many guides have rudimentary English skills and many tourists are not Sinhala or English speaking. The physical interpretation facilities are also inadequate. Many NPs do not have interpretative facilities and in some cases, even if there is a visitor center, the participants do not believe they are functioning properly.

The Travel executive for the Middle East/Pakistan market 1 (personal communication) also noted that some tourists are undertaking a wildlife experience to relax and are not particularly interested in interpretation and education. A number of participants suggested there was a need for research into the requirement of tourists in regards to interpretation.

The need for research into other areas of the industry was also raised. For example, the Owner/Tour executive 5 (personal communication) suggested that there was a need to determine the impacts of tourism on the wildlife and environmental values of the PAs. There was a consensus that the DWLC should take the lead with research. The executive went on to suggest that the lack of research meant that management was guided by research from overseas and that the findings from this research may not be applicable to the situation in Sri Lanka. It was also suggested that better research would make it easier to promote Sri Lankan wildlife products.

Despite the identified importance of research there has been criticism of the research facilities offered by the DWLC. For example, de Silva Wijeyeratne (2006b) noted that "not one of the NPs (administered by the DWLC) of Sri Lanka has a permanent research station welcoming local or foreign researchers. In contrast, hoteliers in Sri Lanka are beginning to offer the use of their hotels as research bases. If this practice continues, the lead for bio-diversity research in Sri Lanka will move out of the DWLC into the domain of the tourism industry."

10.6.11 ENGAGING LOCAL COMMUNITIES

Jayawardena (2013 p 519) argues that by "definition, tourism is about people and their interactions with the natural, social and cultural environments at the places they visit... Yet, tourism is often seen as an elitist business activity with little or no benefit for host communities." He goes on to note that rural communities in Sri Lanka are suspicious of tourism-related developments and that a "key barrier to tourism development in rural areas is the poor image of tourism, which is not perceived to be a safe or respectable career from a rural perspective" (520). This relatively negative image of tourism is unfortunate since, as outlined earlier, wildlife tourism requires the active engagement and support of local communities. It was also noted that despite the negative image amongst some members of the local community many also benefit from the industry. For example, members of the local community are used as guides by the sector. They also provide accommodation and other tourism products (jeep safaris, camping services, and working as trackers) in towns and villages in close proximity and along the popular routes to NPs. For example, it was estimated that more than 200 accommodation units are located in a town neighboring a popular park (Park Manager 11, personal communication). They also supply fruits, vegetables, and other products to the industry. In addition, they also play an important role in on and off-park conservation. Despite the importance of local community members it was acknowledged there were no mechanisms or programs to ensure and deliver benefits from wildlife tourism to the local community. The delivery of benefits was at the discretion of operators.

A number of tourism businesses make concerted efforts to engage with local communities according to some of the interview participants. The Managing Director/Owner Travel Agency 2 (personal communication) remarked that the community received financial and non-financial benefits through tourism and that when his company design wildlife tours they mainly focus on visitor experience and satisfaction as well as giving priorities to providing benefits to the community. Another participant noted that "[O]ur policy is to use the facilities and services of local community wherever we go and operate" (Travel executive 8, personal communication).

One company arranges educational and training programs for local trackers through the Sri Lanka Association for Inbound Tour Operators.

Other participants stated that their companies did not purposely provide any benefits directly to the local community but felt that communities were benefiting anyway from the development of wildlife tourism (Travel executive for Middle East/Pakistan market 1, personal communication). These benefits are particularly important since a number of communities located near NPs face substantial socio-economic problems. Another participant suggested that local communities may benefit indirectly such as from infrastructure development close to the NPs.

Despite the various measures used to reward the local communities, one interview participant argued that tourism companies still do not return enough back to them. He felt that it was particularly important to reward communities in order to address many non-tourism problems facing PAs (Tour executive 3, personal communication).

10.6.12 PROMOTION OF WILDLIFE TOURISM

As noted earlier there is a belief that there is considerable potential for the sector but that it needs greater and better promotion. There was a consensus that effective promotion required a partnership between the public and private sectors but that the government needed to play the leading role. Some felt that the government was not providing sufficient support. There is also an acknowledgment that the wildlife sector has not been developed to a level where Sri Lanka can be marketed as a wildlife destination despite its wildlife assets. There are very popular and competing wildlife destinations globally and currently it is very difficult to promote pure wildlife tourism in Sri Lanka, and as a consequence most of travel agents tend to promote mixed tourism packages/product.

10.6.13 SUGGESTIONS FOR IMPROVEMENT

It is clear that many of the problems facing the wildlife tourism industry are interrelated and require a holistic approach, involving the public and private sectors, that address many of them in conjunction (see Senevirathna & Perera, 2013). The importance of industry involvement was supported by Travel executive 8 (personal communication) who noted that the industry had to take the initiative in protecting wildlife since it is the industry that acquires the major benefits. Travel executive 8 (personal communication) also suggested a need for an integrated regional visitor management strategy that would help relieve the visitor pressure at popular NPs by diverting visitors to other venues. However, Senevirathna and Perera (2013) argue that diversion away from popular parks requires appropriate destination promotion and marketing strategies that highlight the values of the alternative venues. They suggest that the current concentration is largely attributable to a lack of awareness of the unique natural diversity and recreational opportunities available in the range of PAs amongst visitors and tour operators. The ability to provide new opportunities to promote NPs or PAs which are less-known for charismatic species, but contain more subordinate features of biodiversity, will be supported by the fact that nature-based tourism and ecotourism markets in Sri Lanka are becoming more heterogeneous (Perera et al., 2012; Perera & Vlosky, 2013). Another strategy suggested by an interviewee that would help ease congestion at some parks would be to develop various tourism products located on the routes between NPs. This would mean that tour companies would be under less pressure to use NPs located in close proximity to the major tourism hubs. Increased products would allow tour companies to increase the drive time of their tours. It was also suggested that there should be shift in emphasis away from "passive" activities such as jeep safari to the promotion of more "active" experiences such as guided hiking and camping.

Another strategy suggested by de Silva Wijeyeratne (2010) to reduce the overcrowding in Yala is to open up more of the park to tourism. He also suggests that some scrub forest in the park could be can be cleared to create more grassland areas that would result in the congregation of larger numbers of certain species available for viewing by tourists.

Another important initiative suggested for improving management is improved research and the collection of environmental and tourism information. The DWLC are increasing their attention on the concept of carrying capacity in response to concerns over rising visitation in parks and accompanying impacts on resources and on visitor experience (Rathnayake & Gunawardena, 2011). Determining appropriate carrying capacities requires good knowledge of a range of factors. Unfortunately at the present time the managing agencies do not appear to have the funds to undertake this research.

One way of attracting increased funding for research would be to increase park entry fees. As Rathnayake and Gunawardena (2011) suggest, the current fees charged are not indicative of the true value of the product. Increased funds would also allow the DWLC to address a number of other issues as well. For example, improving facilities, the training of DWLC staff, jeep drivers and guides, the enforcement of regulations, and the provision of better interpretation and education all require resources that are clearly not available at the moment. Increased funding could also be used for a formalized program of rewarding the conservation activities of local communities (see Perera & Vlosky, 2013). It appears that currently the delivery of benefits to the local community depends very much on the various tour operators. Addressing these issues would allow Sri Lanka to position itself as a world-class wildlife tourism destination.

10.7 CONCLUSION

Clearly Sri Lanka is in a position to provide visitors with outstanding wildlife tourism experiences however the sector is currently underdeveloped. It is clear that there major problems with overcrowding at some of the more popular NPs as well as a lack of facilities catering for tourists. There are also issues with poorly trained and qualified staff and the provision of inadequate interpretation. The current entry fees into NPs appear to be at the low end of the spectrum and increasing fees would allow park management to address many of the problems associated with the industry. Increasing fees would also allow for local communities to be rewarded for caring for wildlife both on and off-park. The improvement in the industry would also allow for it be more effectively marketed as a wildlife tourism destination.

Interestingly the study participants shared very homogenous views in regards to the industry, its development and the issues it faces. However, it is important to examine the views of other stakeholders including the local community, wildlife tourists, and the Sri Lanka Tourism Development Authority. It is also important to have more research into wildlife tourism occurring in non-natural settings.

KEYWORDS

- **wildlife tourism**
- **Sri Lankan tourism**
- **wildlife conservation**
- **visitor management**
- **protected area management**

REFERENCES

Adams, W. M.; Infield, M. Who is on the Gorilla's Payroll? Claims on Tourist Revenue from a Ugandan National Park. *World Deve*. **2002,** *31* (1), 177–190.

Anon (n.d.). Clearer Skies for Sri Lanka's Hotel Sector, Says RAM Ratings. *The Island Online Website*.[Online], April 2011, Retrieved from 2011 http://www.island.lk/index.php?page_cat=article-detailsandpage=article-detailsandcode_title=24103

Ballantyne, R.; Packer, J.; Sutherland, L. Visitors' Memories of Wildlife Tourism: Implications for the Design of Powerful Interpretive Experiences. *Tour. Manage*. **2011,** *32* (4), 770–779.

Banerjee, A. Is Wildlife Tourism Benefiting Indian Protected Areas? A Survey. *Curr. Issues Tour*. **2012,** *15* (3), 211–227.

British Columbia Ministry of Environment (n.d.); Wildlife Program Plan. British Colombia: British Columbia Ministry of Environment.

Burns, G; Schofield, T. *The Host Community: Social and Cultural Issues Concerning Wildlife Tourism*. Sustainable Tourism CRC: Gold Coast, Queensland, 2001.

Buultjens, J. *In Sri Lankan Tourism: Possibilities for a Sustainable Future.* 7th Tourism Outlook and 3rd Tropical Coastal and Island Tourism Conference, Nature, Culture and Networking for Sustainable Tourism, Dambulla, Sri Lanka, Aug 8–10, 2014.

Buultjens, J.; Ratnayake, I.; Gnanapala, A. Post-Conflict Tourism Development in Sri Lanka: Implications for Building Resilience. *Curr. Issues Tour.* **2015,** DOI: 10.1080/13683500.2014.1002760.

Buultjens, J.; Ratnayake, I.; Gnanapala, A. (2016). Whale watching in Sri Lanka: Perceptions of Sustainability. *Tourism Management Perspectives.* **2016,** *18,* 125–133.

Buultjens, J.; Ratnayake, I.; Gnanapala, A.; Aslam, M. Tourism and its Implications for Management in Ruhuna National Park (Yala), Sri Lanka. *Tour. Manage.* **2005,** *26* (5), 733–742.

Ceylon Tourist Board. *Tourism Management Plan, 1967*. Ceylon Tourist Board: Colombo, 1968.

Charmaz, K. *Contructing Grounded Theory: A Practical Guide through Qualitative Analysis;* Sage Publications Ltd.: London, 2006.

Collaborative Partnership on Sustainable Wildlife Management (CPW) Sustainable Wildlife Management and Biodiversity Fact Sheet, CPW: 2014.

Department of Wildlife Conservation (DWLC). (n.d.); Protected Areas in Sri Lanka. Retrieved June 14, 2014 from http://www.dwc.gov.lk.

de Silva Wijeyeratne, G. Our Big Five. Why Sri Lanka is the Best for Big Game Safaris outside Africa. The Sunday Times Plus. Sunday 18 October 2010. Features. 2010; p 6. Retrieved June 22, 2015 from http://sundaytimes.lk/101017/Plus/plus_23.html

de Silva Wijeyeratne, G. Wildlife Tourism's True Potential. *LMD.* **2006a,** *12* (8), 156. Retrieved June 22, 2015 from http://www.jetwingeco.com/articles/wildlife-tourism%E2%80%99s-true-potential.

de Silva Wijeyeratne, G. Wildlife Tourism Has to Pay Its Way. *LMD.* **2006b,** *12* (11), 60. Retrieved June 22, 2015 from http://www.jetwingeco.com/articles/wildlife-tourism-has-pay-its-way.

Fernando, S.; Bandara, J.; Smith, C. Regaining Missed Opportunities: The Role of Tourism in Post-war Development in Sri Lanka. *Asia Pac. J. Tour. Res.* **2013,** *18* (7), 685–711.

Fernando, S.; Shariff, O. *Ecotourism Faces Serious Challenges in Sri Lanka.* **2013,** Retrieved June 28, 2015 from http://www.tourism-review.com/travel-tourism-magazine-challenges-of-ecotourism-in-sri-lanka-article2113.

Fernandopulle, L. Wildlife Tourism, a Money Spinner - Chitral Jayatilleke, Sunday Observer. **2011,** Retrieved from http://www.sundayobserver.lk/2011/12/04/fin36.asp.

Fredline, L.; Faulkner, B. *International Market Analysis of Wildlife Tourism*, Wildlife Tourism Research Report Series No. 22. CRC for Sustainable Tourism: Gold Coast, Australia, 2001.

Geekiyanage, N.; Vithanage, M.; Wijesekara, H.; Pushpakumara, G. State of the Environment, Environmental Challenges and Governance in Sri Lanka. In *Environmental Challenges and Governance: Diverse Perspectives from Asia;* Mukherjee, S., Chakraborty, D., Eds.; Routledge: Abingdon, Oxon, 2015; pp 92–108. Retrieved June 22, 2015 from https://books.google.com.au/books?id=sPrqBgAAQBAJandpg=PA108andlpg=PA108anddq=FOREST+COVER+REHABILITATION+%E2%80%93+SRI+LANKAandsource=blandots=LDO-3kAMeVandsig=6yroz7qjvmsOjqWGa7Pkh7Mlmb0andhl=enandsa=Xandved=0CDIQ6AEwA2oVChMIl7XuvpfIxwIVRhqmCh0e8gWY#v=onepageandq=FOREST%20COVER%20REHABILITATION%20%E2%80%93%20SRI%20LANKAandf=false.

Gillham, B. *Case Study Research Methods*; Continuum: London, 2000.

Green, R.; Higginbottom, K. *The Negative Effects of Wildlife Tourism on Wildlife;* CRC for Sustainable Tourism: Gold Coast, Australia, 2001.

Higginbottom, K. Wildlife Tourism: An Introduction. In *Wildlife tourism: Impacts, Management and Planning;* Higginbottom, K., Ed.; Common Ground Publishing in Association with the CRC for Sustainable Tourism: Gold Coast, Australia, 2004; pp 1–14.

Higginbottom, K.; Northrope, C.; Green, R. *Positive Effects of Wildlife Tourism on Wildlife*. Wildlife Tourism Research Report Series No. 6. CRC for Sustainable Tourism: Gold Coast, Australia, 2001.

ICRA Lanka 2011; Travel and Tourism: Industry report on Sri Lanka, IMaCS & ICRA: Colombo, Sri Lanka, 2011.

Jayawardena, C. Innovative Solutions for Future Tourism Development in Sri Lanka (2013–2026). *Worldwide Hospitality Tour. Themes.* **2013,** *5* (5), 512–531.

Jayantha, D. (n.d.). *Responsible Elephant Tourism in Sri Lanka.* Retrieved April 2, 2015 from www.european-elephant-group.com/files/.../srilankaelephanttourism.pdf.

Lai, T. Promoting Sustainable Tourism in Sri Lanka. In *Linking Green Productivity to Ecotourism: Experiences in The Asia-Pacific Region;* Hundloe, T., Ed.; Asian Productivity Organization: Tokyo, Japan, 2002; pp 208–214.

Mertens, D. M. *Research Methods in Education and Psychology: Integrating Diversity with Quantitative and Qualitative Approaches* (2nd ed.); Sage: Thousand Oaks, CA, 2005.

Ministry of Economic Development 2011; Tourism Development Strategy 2011–2016. Ministry of Economic Development Sri Lanka: Colombo, Sri Lanka, 2011.

Ministry of Environment and Natural Resources 2002; State of the Environment in Sri Lanka. Ministry of Environment and Natural Resources: Colombo, Sri Lanka, 2002.

Miththapala, S. *Wildlife Tourism in Sri Lanka,* LankaNewspaper. com website. 2012, Retrieved April 10, 2015 from http://www.lankanewspapers.com/news/2012/9/78812_space.html.

Newsome, D.; Dowling, R.; Moore, S. *Wildlife Tourism*; Channel View Publications: Clevedon, UK, 2005.

Patton, M. *Qualitative Evaluation and Research Methods;* SAGE Publications: Beverly Hills, CA, 1990.

Perera, P.; Vlosky, R. P. How Previous Visits Shape Trip Quality, Perceived Value, Satisfaction, and Future Behavioral Intentions: The Case of Forest-Based Ecotourism in Sri Lanka. *IJSMaRT.* **2013**, *11,* 1–24.

Perera, P.; Vlosky, R. P.; Wahala, S. B. Motivational and Behavioral Profiling of Visitors to Forest-based Recreational Destinations in Sri Lanka. *Asia Pac. J. Tour. Res.* **2012,** *17* (4), 451–467.

Ranasinghe, R. Strategic Myopia of Tourism Development in Sri Lanka: A Critique. *Int. J. Multidiscip. Res. Dev.* **2015,** *2* (2), 604–609.

Rathnayake, R. M. W.; Gunawardena, U. A. D. P. Estimation of Recreational Value of Horton Plains National Park in Sri Lanka: A Decision Making Strategy for Natural Resources Management. *J. Trop. For. Environ.* **2011,** *1* (1), 71–86.

Rodger, K.; Smith, A.; Davis, C.; Newsome, D.; Patterson, P. *A Framework to Guide the Sustainability of Wildlife Tourism Operations: Examples of Marine Wildlife Tourism in Western Australia.* CRC for Sustainable Tourism: Gold Coast, Australia, 2010.

Roughguides. (n.d.). *Sri Lanka: National Parks, Reserves and Eco-Tourism.* Retrieved September 15, 2014, from http://www.roughguides.com/destinations/asia/sri-lanka/national-parks-reserves-eco-tourism/#ixzz3jyKQV12P.

Sathurusinghe, A. *Forest Cover Rehabilitation – Sri Lanka, Conservator of Forests;* Forest Department Sri Lanka: Sir Lanka, 2008. Retrieved June 12, 2015 fromwww.iufro.org%2Fdownload%2Ffile%2F7402%2F5122%2FSri_Lanka_pdf%2Fandei=iHHeVZfRCMa0mAWe5JfACQandusg=AFQjCNEppJ2wbIam30Jt__Fj5F_s_eTYogandsig2=bjb3e8y4EUdmzRYGiIat6g.

Senevirathna, H.; Perera, P. Wildlife Viewing Preferences of Visitors to Sri Lanka's National Parks: Implications for Visitor Management and Sustainable Tourism Planning. *JTFE.* **2013,** *3* (2), 1–10.

Shackley, M. *Wildlife Tourism*; International Thomson Business Press: London, 1996.

Sharpley, R. Tourism and Sustainable Development: Exploring the Theoretical Divide. *J. Sust. Tour.* **2000,** *8* (1), 1–19.

Sri Lankan Tourism Development Authority (n.d.); Tourism for All: National Strategy for Sri Lanka Tourism 2009–2012. Retrieved September 10, 2012 from the Sri Lankan Tourism Development Authority Website: http://www.sltda.lk/.

Sri Lanka Tourism Development Authority 2014; Annual Statistical Report – 2014. Sri Lanka Tourism Development Authority: Colombo, Sri Lanka, 2014.

Sri Lanka Tourism Development Authority 2013; Survey of Departing Foreign Tourists from Sri Lanka 2013. Sri Lanka Tourism Development Authority: Colombo, Sri Lanka, 2013.

Sri Lanka Tourism Development Authority 2009; Annual Statistical Report - 2009. Sri Lanka Tourism Development Authority: Colombo, Sri Lanka, 2009.

Sri Lanka Tourism Development Authority 2005; Annual Statistical Report - 2005. Sri Lanka Tourism Development Authority: Colombo, Sri Lanka, 2005.

STCRC 2009; Wildlife Tourism: Challenges, Opportunities and Managing the Future. Sustainable Tourism CRC: Gold Coast, Australia, 2009.

Tisdell, C.; Wilson, C. *Nature-based Tourism and Conservation: New Economic Insights and Case Studies;* Edward Elgar Publishing: Gloucester, England, 2012.

Tisdell, C.; Wilson, C. Do Open-Cycle Hatcheries Relying on Tourism Conserve Sea Turtles? Sri Lankan Developments and Economic–Ecological Considerations. *Environ. Manage*. **2005a,** *35* (4), 441–452.

Tisdell, C.; Wilson, C. Perceived Impacts of Ecotourism on Environmental Learning and Conservation: Turtle Watching as a Case Study. *Environ. Deve. Sust*. **2005b,** *7,* 291–302.

Tisdell, C.; Wilson, C. *Economics, Wildlife Tourism and Conservation: Three Case Studies;* Sustainable Tourism CRC: Gold Coast, Australia, 2004.

Udaya, S. N. Local People's Attitudes towards Conservation and Wildlife Tourism around Sariska Tiger Reserve. *Indian J. Environ. Manage*. **2003,** *69* (4), 339–347.

Wilson, C.; Tisdell, C. Sea Turtles as a Non-Consumptive Tourism Resource Especially in Australia. *Tour. Manage*. **2001,** *22,* 279–288.

World Bank (n.d.). *Terrestrial Protected Areas (% of Total Land Area)*. Retrieved August 26, 2015 from http://data.worldbank.org/indicator/ER.LND.PTLD.ZS

CHAPTER 11

STATUS AND PROSPECTS FOR THE FUTURE OF WILDLIFE TOURISM IN BANGLADESH

MD. GOLAM RABBI*

Wildlife & Nature Conservation Circle, Forest Department, Bangladesh

E-mail: rabbi_rk@yahoo.com

CONTENTS

ABSTRACT

Geographically Bangladesh is the utmost destination of tourism for its numerous panoramic beauties. Tourist attractions include historical monuments, world longest sandy sea beach, diverse forest ecosystems, wildlife of various species, the Sundarbans, and tribal people. This is a land of eco-tourism and wildlife research. The Bengal Tiger and its unique mangrove habitat, the Sundarbans is the most tourist hub in Bangladesh. Bangladesh is also home to other well-known wildlife including Asian Elephant, Hoolock Gibbon, Asian Black Bear, Spotted Deer, Saltwater Crocodile, King Cobra, Python, and so forth. Bangladesh supports 130 species of mammals, 710 species of birds (310 species of migratory birds), 164 species of reptiles, and 56 species of amphibians. A total of 219 species including fishes, amphibians, reptiles, birds, and mammals are being noted as threatened according to Bangladesh National Criteria. Notable wildlife species that have already extinct from Bangladesh are the one and two horned Rhinoceros, Gaur, Banteng, Swamp Deer, Nilgai, Wolf, wild Water Buffalo, Marsh Crocodile, Common Peafowl, and Pink Headed Duck. To conserve the wildlife and its habitat, government of Bangladesh has declared 38 Protected Areas (PAs) including 17 National Parks, 20 Wildlife Sanctuaries. Besides this, government has also notified one Marine PA (Swatch of no Ground) for the conservation of aquatic mammals and fisheries; two Vulture Save Zones exclusively for threatened vulture species. These PAs are distributed into the four major forest typologies of Bangladesh, Tropical Evergreen and Semi Evergreen Forest (Hill Forest), Moist Deciduous Forest, Mangrove Forest, and Coastal Forest. The rich biodiversity is significant to the local context as well as to the international context. Two Ramsar sites Tanguar Haor and three Wildlife Sanctuaries of the Sundarbans (East, West, and South) in 1992 have been declared in Bangladesh and the entire Sundarbans (world's largest mangrove tract) was being designated as a World Heritage Site in 1997.

11.1 INTRODUCTION

The People's Republic of Bangladesh is situated in the northeastern part of the South Asian subcontinent, situated between 20° 25′ and 26° 38′ north latitude and 88° 01′ and 92° 40′ east longitude. Total area of this country is about 147,570 km^2 with three broad physiographic regions, flood plains, terraces, hills, and so forth occupying about 80, 8, and about 12% of the total land area respectively. The majority parts of Bangladesh are less than 12 m (39 ft) above the sea level, and the mountainous regions on the north-east and southeast exist with an average altitude of 244 and 610 m, respectively. The peak point of the state (1230 m) is located at the southeastern edge of the erstwhile district of Chittagong Hill Tracts. Till today Government of Bangladesh has declared 39 Protected Areas (PAs) including 17 National Parks, 20 Wildlife Sanctuaries, 1 Marine PA, 1 Special Biodiversity Conservation Area, and also declared 2 Vulture Save Zones. Forest PA covers 439,881.31 ha excluding Marine PA and Vultures Saves Zones and represents 2.98% of total area of the country. These PAs are distributed into the four major forest typologies of Bangladesh, namely Tropical Evergreen and Semi Evergreen Forest (or Hill Forest), Moist Deciduous Forest, Mangrove Forest, and Coastal Forest.

Bio-geographically, Bangladesh situated at the junction of the Indian and Malayan sub-regions of the Indo-Malayan Realm and is placed very near to the western side of Sino-Japanese region. The country's biodiversity imitates this mixture. Huge number of native floras, including 3000–4000 species of woody floras, have been recorded from Bangladesh, and also supports 130 species of mammals, 710 species of birds, 164 species of reptiles, and 56 species of amphibians (Khan, 2013) (Table 11.1). Rich aquatic biodiversity comprises 260 species of finfish belonging to 56 families, 43 species of freshwater and land Mollusks, 246 bryophytes species, 195 species of pteridophytes, and 429 species of butterflies. Bangladesh's aquatic diversity, especially marine ecology, has not yet been adequately discovered. A total of 219 species including fishes, amphibians, reptiles, birds, and mammals are being noted as threatened according to Bangladesh National Criteria (IUCN Bangladesh, 2006). The rich biodiversity is significant to the local context as well as to the international context. Two Ramsar sites Tanguar Haor and three Wildlife Sanctuaries of the Sundarbans (East, West, and South) in 1992 have been declared in Bangladesh and the entire Sundarbans (world's largest mangrove tract) was being designated as a World Heritage Site in 1997 (Table 11.2).

TABLE 11.1 Present Conservation Status of Wildlife in Bangladesh.

Animals group (terrestrial and marine)	No. of species of resident wild animals (terrestrial and marine)	No. of species extinct from the nature	No. of threatened and endangered species				Data deficient	No. of species not threatened
			Critically endangered	Endangered	Vulnerable	Total		
Fish	708	0	12	29	17	58	66	584
Amphibian	56	0	0	3	5	8	7	7
Reptile	164	1	13	28	22	63	39	24
Bird	710	2	19	20	8	47	164	407
Mammal	130	10	21	15	7	43	53	17
Total	1644	13	65	95	59	219	329	1039

*Source: IUCN 2006 and Khan 2013.

TABLE 11.2 Protected Areas: List of Notified Protected Areas (PAs) of Bangladesh.

National Park

Sl. no.	National park	Location	Area (ha)	Date of notification
1.	Bhawal National Park	Gazipur	5022.29	11-05-1982
2.	Modhupur National Park	Tangail/ Mynsingh	8436.13	24-02-1982
3.	Ramsagar National Park	Dinajpur	27.75	30-04-2001
4.	Himchari National Park	Cox's Bazar	1729.00	15-02-1980
5.	Lawachara National Park	Moulavibazar	1250.00	07-07-1996
6.	Kaptai National Park	Ctg. Hill Tracts	5464.78	09-09-1999
7.	Nijhum Dweep National Park	Noakhali	16,352.23	08-04-2001
8.	Medha Kassapia National Park	Cox's Bazar	395.92	04-04-2004
9.	Satchari National Park	Habigonj	242.91	15-10-2005
10.	Khadeem Nagar National Park	Sylhet	678.80	13-04-2006
11.	Baraiyadhala National Park	Chittagong	2933.61	06-04-2010
12.	Kadigar National Park	Mymensing	344.13	24-10-2010
13.	Shingra National Park	Dinajpur	305.69	24-10-2010
14.	Nababgong National Park	Dinajpur	517.61	24-10-2010
15.	Kuakata National Park	Patuakhali	1613.00	24-10-2010
16.	Altadeghe National Park	Nagaon	264.12	14-12-2011
17.	Birgonj National Park	Dinajpur	168.56	14-12-2011

Wildlife Sanctuary

Sl. no.	Wildlife Sanctuary	Location	Area (ha)	Date of notification
18.	Rema-kelenga Wildlife Sanctuary	Hobigonj	1795.54	07-07-1996
19.	Char Kukri-Mukri Wildlife Sanctuary	Bhola	40.00	19-12-1981
20.	Sundarban (East) Wildlife Sanctuary	Bagerhat	31,226.94	06-04-1996
21.	Sundarban (West) Wildlife Sanctuary	Satkhira	71,502.10	06-04-1996
22.	Sundarban (South) Wildlife Sanctuary	Khulna	36,970.45	06-04-1996
23.	Pablakhali Wildlife Sanctuary	Ctg. Hill Tracts	42,069.37	20-09-1983
24.	Chunati Wildlife Sanctuary	Chittagong	7763.97	18-03-1986
25.	Fashiakhali Wildlife Sanctuary	Cox's Bazar	1302.42	11-04-2007
26.	Dudh Pukuria-Dhopachari Wildlife Sanctuary	Chittagong	4716.57	06-04-2010

TABLE 11.2 *(Continued)*

Sl. no.	Wildlife Sanctuary	Location	Area (ha)	Date of notification
27.	Hazarikhil Wildlife Sanctuary	Chittagong	1177.53	06-04-2010
28.	Shangu Wildlife Sanctuary	Bandarban	2331.98	06-04-2010
29.	Teknaf Wildlife Sanctuary	Cox's Bazar	11,614.57	24-03-2010
30.	Tengragree Wildlife Sanctuary	Barguna	4048.58	24-10-2010
31.	Sonarchar Wildlife Sanctuary	Patuakhali	2026.48	24-12-2011
32.	Chandpai Wildlife Sanctuary	Bagherhat	560.00	29-01-2012
33.	Dudmukhi Wildlife Sanctuary	Bagherhat	170.00	29-01-2012
34.	Daingmari Wildlife Sanctuary	Bagherhat	340.00	29-01-2012
35.	Nagarbari-Mohangonj Dolphin (Platanista gangetica) Sanctuary	Pabna	408.11	01-12-2013
36.	Shilanda-Nagdemra Wildlilfe (Dolphin) Sanctuary	Pabna	24.17	01-12-2013
37.	Nazirgonj Wildlilfe (Dolphin) Sanctuary	Pabna	146.00	01-12-2013
	Sub Total		220,334.78	

Marine Protected Area

Sl. no.	Marine Protected Area	Location	Area (ha)	Date of notification
38.	Swatch of No-ground Marine Protected Area	South Bay of Bengal	173,800.00	27-10-2014

Vulture Safe Zone

Sl. no.	Vulture Safe Zone	Location	Area (ha)	Date of notification
39.	Vulture Save Zone	Shylet, Hobigonj, Sunamgonj, Molvibazar, Netrokona, Kishorgonj, Gazipur, Mymanshing, Brammonbaria, Norshindi, Comilla, and Khagrachari	1,966,300.18	23-12-2014
40.	Vulture Save Zone	Faridpur, Magura, Jhinaidha, Madaripur, Jossore, Gopalgonj (Except Tungipara), Narial, Shariatpur, Barishal, Bagerhat, Khulna, Satkhira, Pirojpur (Except Vandaria), Jhalakhati, Patuakhali, and Barguna.	2,771,700.26	23-12-2014
		Sub-Total	4,738,000.44	

TABLE 11.2 *(Continued)*

Special Biodiversity Conservation Area				
Sl. no.	**Special Biodiversity Conservation Area**	**Location**	**Area (ha)**	**Date of notification**
41.	Ratar Gul	Sylhet	204.25	31-05-2015

*Source: Bangladesh Forest Department.
**Though Vulture Save Zones are declared under Wildlife (conservation & security) Act, 2012 but it was not notified as Protected Area because most of the areas belong to private property.
***Marine Protected Area is the outer side of the Sundarbans and have common border with India.

11.2 WILDLIFE TOURIST HOT SPOTS IN BANGLADESH

11.2.1 BHAWAL NATIONAL PARK

Bhawal National Park was established and preserved as a National Park in 1974. It was declared officially in 1982. It is located in Gazipur district, Dhaka Division of Bangladesh. Its principle is to protect important habitats as well as to afford opportunities for recreation. It belongs to the International Union for Conservation of Nature (IUCN) Management Category V, as a protected landscape. It is a tropical moist deciduous forest, where Sal is the main tree species and locally is called "Gajari forests." This Sal forest is a secondary forest originated from coppice. The area was renowned for Peacocks, Tiger, Elephant, Leopard, Clouded Leopard, Black Panther, and Sambar Deer. However, much of the wildlife had been disappeared and only a few species remain. Approximately 18 species of mammals, 11 species of amphibians, 19 species of reptiles, and 84 species of birds are found in the park area. Civet, Mongoose, Fox, Jungle Cat, Wild Boar, and Hare are the main mammals. Monitor lizard, Snake, Python, and Tortoise are the main reptiles. Most of this region was covered by forests 50 years ago and the dominant species was Sal (*Shorea robusta*). A total of 390 species of plants are being recognized in the PA area among which are 35 species of climbers, 127 species of herbs, 60 species of shrubs, and 168 species of timber trees. This is the most popular tourist spot near the Dhaka City. The park is nearly 80 km away from the capital and well communicated by bus.

11.2.2 MADHUPUR NATIONAL PARK

The Madhupur national Park is situated in the northern part of Bhawal-Madhupur Shal (*Shorea robusta*) forest tract, somewhat 50 km south of the Garo Hills of the Meghalaya State of India, and about 151 km north of Dhaka, Bangladesh. The Park is deciduous with a slight mixture of evergreen forest. The main plant species of the forest is Shal *(Shorea robusta*). In Modhupur National Park, the total number of identified plant species is 176. There are four Rubber plantations surrounding the site, namely Pirgacha Rubber Garden, Chandpur Rubber Garden, Sontoshpur Rubber Garden, and Kamalapur Rubber Garden. This park also has wide variety of fauna. Identified fauna species include 4 amphibians, 7 reptiles, 11 mammals, and 38 birds species. This deep forest is the key place for both researcher and tourist.

11.2.3 SATCHARI NATIONAL PARK

Satchari National Park (SNP) to begin with supported an indigenous vegetation of mixed tropical evergreen forest. On the other hand, almost all of the original forest has been removed or substantially altered, turning it into a secondary forest. Amongst the wildlife found in SNP are 6 species of amphibians, 18 species of reptiles, 149 species of birds, and 24 species of mammals. Hoolock Gibbon and Phayre's Langur are the keystone species of SNP with lesser-known species like the Htun Win's Tree Frog have only been recently recognized. Among the bird species Oriental Pied Hornbill, Red Jungle Fowl, and Red headed Trogon are common. A number of wildlife species have gradually become extinct from the park like the Tiger, Barking Deer, Leopard, Porcupine, and Wild Cow to mention a few. It is paradise for bird lovers.

11.2.4 LAWACHARA NATIONAL PARK

Lawachara National Park (Lawachara NP) is a nature reserve in Bangladesh situated at Maulvibazar District in the northeastern region of the country. The area was declared as national park by the government on July 7, 1996. The Lawachara NP is a mixed tropical evergreen forests and a biodiversity hotspot in the country with many endangered and endemic species of fauna and flora (Mukul, 2008, 2014). The highly productive ecosystem of the Lawachara NP provides a wide range of valuable services, for example, about 167 species of plants, 276 species of wildlife, including 246 species of birds, 20 species of mammals, 6 species of reptiles, 4 species of amphibians as well as the flagship species such as the critically endangered Hoolock Gibbons of which merely 62 individuals remain in the area (Leech & Ali, 1997; FSP, 2000b; Thompson & Johnson, 1999, 2003; Feeroz & Islam, 2000; Ahsan, 1995a; Khan, 1982). In order to enhance economic benefit plantation program in the forest was started as early as 1923 with establishment of reserve forest. Since then plantation took place till 1984 (FD, 1997). A large number of tourists visit particularly easily accessible parts of Lawachara to have a feel of luxuriant vegetation of evergreen forests and good landscape of the Park with rolling hills and interspersed valleys. An estimate by Feeroz and Islam (2000) illustrates that during December–February, in average, 300 buses and minibuses/month visit the park. According to them, the visitors stay throughout the day and leave in the evening. According to Ahsan (1995b), during winter, every day 2–5 groups of visitors come to the park by buses and microbuses, comprising about 15–100 people/group. The place is well developed with tourist facilities including tea garden lodge, hill view eco-cottage, five star hotels, and so forth.

11.2.5 HIMCHARI NATIONAL PARK

Himchari National Park is just south of the town of Cox's Bazar which consist of abundant tropical rain forest, grasslands, and trees, and features a numeral of waterfalls, the largest of which cascades downward to the sandy, sun-drenched beach. The flora and fauna in the part is rich and birders in no way fail to be delighted at the extensive birdlife. At once the stomping ground was herds of Asian Elephant. Himchari is still habitat to a small number of these majestic animals. Other mammals like Gibbon, Rhesus Macaque, Leopard Cat, Fishing Cat, Sloth Bear, Wild Boar, Dhole (also known as the Asiatic Wild Dog), and Indian Muntjac are present in the park. Additionally the Himchari National Park is home to around 56 species of reptiles, 55 species mammals, and 14 amphibian species. There are about 100 species of trees, shrubs, canes, palms, ferns, grasses, and herbs in the park. This extensive diversity of plant life provides appropriate habitats for more than 286 species of birds that call the park home, including the Barn Swallow, Asian Palm Swift, and Jungle Myna. The most astonishing thing of this park is the seashore of the Bay of Bengal with natural forest on the other side.

11.2.6 KAPTAI NATIONAL PARK

The semi-ever green forest encompasses very high hills, springs, canals, and rivers. Kaptai National Park is located in the Rangamati Hill Tracts, which cascades between the Karnaphuly and Kaptai Mountain Ranges. It is an extremely astonishing natural area that has to be experienced to be truly esteemed. This is where undulating green hills greet the spectacular blue azure of the Kaptai Lake in an exciting combination of color. Various natural springs each add their own unique appeal to the park while wildlife stun the public as they scuttle about on their regular activities. Some of the wildlife that inhabits the park includes Elephant, Deer, Jungle Cat, Monkey, and so forth. Almost 60% of wildlife taking place in Bangladesh found here. The tropical rain forest that is originated on the banks of the Karnaphuly River is another amazing feature of this lovely park. Every year millions of national and foreign tourists come to this place to see the beautiful scenario.

11.2.7 TEKNAF WILDLIFE SANCTUARY

Located on the banks of the Naf River in Teknaf, Cox's Bazaar and occupied almost 11,500 ha. A highlight of the Teknaf Wildlife Sanctuary is the Kudum Cave, more usually referred to as the "Bat Cave" for noticeable reasons. The two species of bats, of which there a large number, are not the only cave inhabitants. Kudum Cave is also habitat to four species of snails, four species of fish dwelling in the underwater pools, and three species of spiders. The sanctuary harbors a wide diversity of tropical semi-evergreen flora and fauna. The Sanctuary contains 55 mammals like the Slow Bengal Loris and the Masked Palm Civet; 286 birds including the White-Bellied Seagull, White-Browed Piculet along with some Kingfisher and Bittern species; 56 reptiles, 13 amphibians, and 290 species of plants. Currently, Asian Elephants, Wild Boars, Clouded Leopards, Hog-badgers, Deer, and so forth are found in plenty. Only way to this place is by bus with accommodation facilities are Forest rest house, hotels, and eco-cottages in Teknaf.

11.2.8 REMA-KALENGA WILDLIFE SANCTUARY

This is a dry and evergreen forest. It is also the second largest natural forest in Bangladesh after the Sundarbans. It is the country's second largest wildlife sanctuary and the most prosperous in terms of biodiversity. It is situated in Habiganj district. It is a pleasant combination of diverse kinds of plants', animals', and birds' life. Tourist can get one of the best experiences of bird watching here. A total of 167 species of bird's inhabitants here, for example, Greater Yellownape, Asian Barred Owlet, and Great Racket-Tailed Drongo. Primates included Hoolock Gibbon, Phayres Langur, Rhesus Monkey, Capped Languer, Slow Loris, Fishing Cat, Wild

Boar, and Barking Deer are in this forest. Identified species includes 37 mammals, 7 amphibians, 18 reptiles, and 638 plants are in this forest. The prime communication way to the park is by bus or car.

11.2.9 PABLAKHALI WILDLIFE SANCTUARY

Pablakhali Wildlife sanctuary is the optimum hill forest left over in Bangladesh and an amazing wetland site also. Main attraction of this place is Asiatic Elephant and a lot of species of native birds. Other mammals are Hoolock Gibbon, Capped Langur, Rhesus Macaque, Dhole, Small Cats, Otters, Wild Boar, and Sambar. Although 183 bird species, 76 mammal species, and many other species of reptiles and amphibians were recorded. In addition to these animals, a large variety of Snakes, Lizards, and other reptiles are also very common to this area. The list of wild birds is also quite long, notable species being Pigeons, Doves, Jungle Fowl, Partridge, Pheasants, Mayna, Woodpecker, Cuckoo, Owl, Adjutant, Thrush, Babbler, Drongo, Grackle, Chat Robin, Swallow, Bee-eater, Hoopee, Teals, Quails, Wild Ducks, and so forth. There are two rest houses available for the tourists.

11.2.10 CHUNATI WILDLIFE SANCTUARY

Chunati Wildlife Sanctuary is about 70 km south of Chittagong city and on the way of Cox's bazaar highway. This is amazing and rare kind of ecosystem in Bangladesh. The Sanctuary is also declared as MIKE site. South of the forest is still roaming area of Asian Elephant. This area is continuous natural panorama of hills and grasslands. Rhesus Macaque, Barking Deer included with 19 mammals, 53 birds, 4 amphibians, and 7 reptiles in this forest. The natural trails of the Sanctuary are ideal place for hiking. It is a great resource of eco-tourism. Rest houses are obtainable for tourists manage by the Forest Department.

11.2.11 DUDHPUKURIA-DHOPACHARI WILDLIFE SANCTUARY

Dudhpukuria-Dhopachari Wildlife Sanctuary is tropical evergreen and semi-evergreen forest located in Rangunia, Chittagong. The Wildlife Sanctuary contains significantly high floral and faunal diversities. Garjan is the dominant tree species along with its correlate plant species. Asian Elephant, Gibbon, Wild Boar, Rhesus Macaque, and Barking Deer are the most prominent mammal among the faunal species. Forest rest house is obtainable for the tourist.

11.2.12 KUAKATA NATIONAL PARK

It is situated in Potuakhali district. The park, located on the seashore, encompass costal mangrove plantations. Main attraction of the park is long sandy seashore. Besides this the sandy beach is the egg laying ground for thousands of marine turtle. The park also supports a lot of wildlife. Keora and Bain are the main plant species of this park, a mixture of plant-based resources such as Reeds, Catkin, Grass, Hental, Helipana, and Maila are found in plenty throughout the area. Kuakata is one of the most attracting sightseer places of the country. The foremost way to go the site is by bus and lots of standard hotels, motels, and eco-cottages are accessible.

11.2.13 THE SUNDARBANS

The Sundarbans is the world's largest coastal wetland shared between Bangladesh (62%) and India (38%). In the line with the bio-geographical zoning approach, five habitat typologies are recognized namely; Shore, Low Mangrove Forest, High Mangrove Forest, Open Land or Grassland,

and Estuarine River Base. The Shore habitat covers from the open sandy to muddy areas along the edges of the Wildlife Sanctuary on the Bay of Bengal which generally serves as the major habitat of distinctive bird species in the Sundarbans. The seashore is rich in minuscule aquatic organisms including shells, crabs, shrimps, and so forth. The Sundarbans was declared as *"World Heritage Site"* in 1997. In addition, three Wildlife Sanctuaries, Government of Bangladesh has also declared three new wildlife (dolphin) sanctuaries in 2013 along the shore of the Sundarbans to conserve large number of Ganges River Dolphins or Shushuks, Irrawaddy Dolphins, and Finless Porpoises.

Low Mangrove habitat is a tidal area and usually characterized by low undergrowth composed of small trees, shrubs, hentals, and others. The Sundarbans is eminent for the Charismatic and Flagship species the Royal Bengal Tiger. It is the merely mangrove habitat in the world for the Royal Bengal Tiger, *Panthera tigris tigris* species. According to the camera-trapping census 2015 in the Sundarbans only 106 tigers left. It harbors other important wildlife, for example, globally threatened Estuarine Crocodile, Indian Python, Ganges and Irrawaddy Dolphins, and the critically endangered endemic river Terrapin (*Batagur baska*). Here tourists also catch the attention of other wildlife, for example Spotted Deer, Wild Boar, Rhesus Monkey, Otter, Jackal, and so forth.

Mangrove Forest is generally characterized by high vegetation consisting of medium to large trees such as Sundri, Gewa, Keora, Baen, Passur, and others. Its special biodiversity is expressed in a large assortment of flora; 334 plant species belongs to 245 genera and 75 families, 165 algae, and 13 orchid species. It is also rich in faunal diversity with 693 species of wildlife which includes; 49 mammals, 59 reptiles, 9 amphibians, 210 white fishes, 26 shrimps, 14 crabs, and 43 mollusks species. The varied and colorful bird-life found along the channel of the property is one of its greatest attractions including 315 species of waterfowl, raptors, and native birds including nine species of kingfisher and the magnificent white-bellied sea eagle.

The prime way to visit the Sundarbans is by water vessel. But before that tourists have to go Khulna by bus, by train, or by air up to Jessore and then Khulna by bus or car. There are some authorized travel agencies have package for tourist in the winter season including night halt on vessel inside the Sundarbans. This is the great opportunity for all nature devotees to enjoy the jungle safari with the Royal Bengal Tiger.

11.2.14 NIJHUM DWIP NATIONAL PARK

Nijhum Dwip Island is a small island under Hatiya upazila of Noakhali District in Bangladesh. The Island is the southernmost island of Bangladesh. A cluster of islands mainly; Ballar Char, Char Osman, Kamlar Char, and Char Muri emerged in the early on 1950s as an alluvium in the shallow estuary of the Bay of Bengal On the south of Noakhali. One of the key shorebird sites in the East Asian–Australasian Flyways. Huge numbers of migratory shore birds visit the islands, such as Waders, Gulls, Terns, Egrets, Ducks, Geese, and so forth. Some rare species like Spotted Green Shank, Spotted Red Shank, Spoon-billed Sandpiper, Indian Skimmer, Sandpiper,

Wagtail, and Brown-headed Gull are also being seen. Among the trees Keora is common everywhere. In addition Gewa, Kankra, Bain, Babul, Karamja, Pashur, and many other species also found in this island. The main attraction in these forests is the herd of about 20,000 Spotted Deer (*Axis axis*) (Rabbi, 2011). Some private eco-cottages and hotels are in the Island and Hatia. Accommodation water vessels are available from Dhaka to Hatia.

11.2.15 SONADIA ISLAND

Sonadia Island is located in Cox's Bazaar with an area of 4916ha. The Island was declared an Ecologically Critical Area (ECA) by the Government of Bangladesh in 1999. Tourist and researchers have been fascinated mostly for mangrove, marine turtles, and migratory bird watching. The island is prosperous in species despite its small size. The significant habitats at the site include mangrove, mudflats, beaches and sand dunes, canals and lagoons, and marine habitat. The place supports the very last enduring remnant of natural Mangrove Forest in southeast Bangladesh. The site lies on the East Asian Australasian and Central Asian Flyways and the mangrove and shallow shoals surrounding the island offer an excellent wintering area for migratory waterfowl and shorebirds, including three globally threatened species Spoon-bill Sandpiper, Nordmann's Greenshank, and Great Knot. The sandy beaches and sand dunes support some globally threatened marine turtle species. Other important species include Irrawaddy dolphin and crustacean species, a wild grass relative of rice, fishes, and mollusks. Bus and airline services and are available from Dhaka to Cox's Bazar. Then speed boat needs to be hired for Sonadia Island. There is no accommodation in Sonadia. So visitors have to stay at Cox's Bazar.

11.2.16 CHAR KUKRI MUKRI WILDLIFE SANCTUARY

Char Kukri Mukri Wildlife Sanctuary is an offshore island of Char Fasson under Bhola district. The area of this sanctuary is around 40ha, which is alienated from Bhola mainland by the river Meghna. Composition of coastal mangrove is dissimilar than that of the Sundarbans. Keora (*Sonneretia apetala*) is prevalent and Gewa (*Gumlina areborea*) is also present. Among the wildlife Spotted Deer, Monkey, and Wild Boars are common. Mammals also include Fishing Cat and Oriental Small-clawed Otter. More than eight species of Herons breed in the Sanctuary. Other waterfowls include Egrets, Bitterns, and Grey Pelican. All three of the Monitor species dwelling in Bangladesh were being reported from this Sanctuary. The only way to this place is by water vessel from Dhaka (Sadarghat Launch Terminal).

11.2.17 TANGUAR HAOR

Tanguar Haor is a sole wetland ecosystem of national and international significance. The Haor address to swamp forests, hundreds of bird's species, fishes, amphibians, reptiles, and mammals, many of which are rare and endangered. Tanguar Hoar is roughly 10,000 ha and is located in Sunamgonj, northeastern district of Bangladesh. It was declared an ECA in 1999 by the Government of Bangladesh, and a "Ramsar Site," wetland of national and international importance, by the Ramsar Bureau in 2000. The wetland supports freshwater mother fisheries and is a recognized residence in every winter to nearly 200 migratory bird species. The prime way to the Haor is by bus or train. Tourists can enjoy the beautiful scenario with amazing food staying in eco-cottage managed by stakeholders. Non-governmental organizations (NGOs) also have some cottage to facilitate tourists.

11.2.18 HAIL HAOR

Hail Haor is a basin between hills that becomes a large solitary body of water which overall catchments area 60,000 ha, Wet Season Haor area

12,490 ha, Dry Season water area 4009 ha (March 2000), 400 ha (1999 dry season), and Adjacent floodplain 20,000+ ha. Out of total area, 100 ha were declared as a Fish Sanctuary known as Baikka Beel. Hail Haor comprises of open water with emergent vegetation (mostly lotus), and a fringe of resident swamp forest planted about 10 years ago. Initially it was protected to conserve and restore fish and it also supports about 90 species of fish, but populations of wintering water birds have increased after the declaration of Fish Sanctuary. So far 141 bird species have been documented within the sanctuary. It habitually supports nearly 20,000 migratory water birds every year. Up to nine wintering Pallas's Fish Eagles, large flocks of ducks including up to 4500 Fulvous Whistling Duck (*Dendrocygna bicolor*) occur in winter, Greater Spotted Eagle is regular, and good numbers and diversity of shorebirds, marshland warblers, and other birds occur. This is the only substantial community managed wetland sanctuary in the country.

Due to its picturesque attractiveness and scenery, it attracts huge number of tourists to visit Baikka Beel every year mostly in the winter season. The beel is very much closer to Srimangal, Moulovibazar and well communicated both by bus and train. Accommodation facilities are also well developed by NGOs and Local community.

11.2.19 BSM SAFARI PARK, GAZIPUR

Bangabandhu Sheikh Mujib Safari Park (BSMSP) is one of the best tourist attractions in Bangladesh. Tourist will get the flavor of visiting wild Africa here. It is innovative and attractive visiting place. The Park has established with an area of 3690 acres. On the basis of actual field situation including 1335 acres Core Safari Park, 576 acres Safari Kingdom, 955 acres Biodiversity Park, and 824 acres Extensive Asian Core Safari Park.

The Natural forest of BSMSP is vigorous plant storage. All group of plants like trees shrubs, herbs, climbers which are gymnosperm, angiosperm, pteridophyte, bryophyte, epiphyte, and parasite, and so forth are available in natural forest of this park. Besides different type of ornamental plants, orchids, seasonal flowering plants, medicinal plants, and fruit bearing plants are also present here. There are a total number of 276 plant species under 227 genera and 80 families in the natural forest of this park. The park is home for both wild and cage animals. In wild condition the park supports about 78 species of birds, 5 amphibians, 14 mammals, and 13 reptiles. The main attraction of the park is cage animals for the tourist. About 5000 visitors visit the park. The park becomes the main tourist hotspot of the capital dwellers. Notable cage wildlife's are Royal Bengal Tiger, Asiatic Lion, White Lion, Asiatic Black Bear, Giraffe, Barking Deer, Gharial, Salt water Crocodile, Turtle, Ostrich, Emu, Great Hornbill, Pelican, and Flamingo. A butterfly garden is also established including about 50 butterfly species. The major communication way to this park is bus, car, and so forth. It will take 2 h from the capital.

11.3 CONCLUSION

Bangladesh is a land of wildlife, hills and forests, sandy sea beach, and other panoramic beauties. Here tourist can hear the roar of the Royal Bengal Tiger in the largest Mangrove Forest, the Sundarbans. Bangladesh Government has taken many projects to make sure the best facilities for both national and international nature lovers.

Heartily welcome to "Beautiful Bangladesh."

11.4 ACKNOWLEDGMENT

Some information was collected from various sources notable; Bangladesh Forest Department, IUCN Bangladesh, Nishorgo Support Project (NSP), BSM Safari Park Project, Internet Articles, and so forth. Special thanks to Dr. Tapan Kumar Dey and Mr. Hoq Mahbub Morshed for sharing ideas and views. Few photo credits go to Dr. Bitapi C. Sinha, Mr. Ashit Ranjan Paul, IUCN Bangladesh, Shimanto Dipu, Wildlife Center, Md. Faysal Ahmad, A.E.M. Rubayet Elahi, Md. Shohel Rana, and Israt Jahan.

KEYWORDS

- **the Sundarbans**
- **Mangrove Forest**
- **Bengal Tiger**
- **Asian Elephant**
- **Hoolock Gibbon**
- **Swatch of no Ground**

REFERENCES

Ahsan, M. F. Fighting Between Two Females for a Male in the Hoolock Gibbon. *Int. J. Primatol.* **1995a**, *16* (5), 131–137.

Ahsan, M. F. *Human Impact on 2 Forests of Bangladesh: A Preliminary Case Study*. International Wildlife Management Congress, The Wildlife Society: Bethesda, MD, 1995b; pp 368–372.

Bangladesh Forest Department (BFD); Ministry of Environment and Forest, Government of Bangladesh. (http://www.bforest.gov.bd).

FD (Forest Department)1997; Sylhet Forest Division -at a Glance. Forest Department: Bangladesh, 1997, p 34.

Feeroz, M. M.; Islam, M. A. Primates of the West Bhanugach Forest Reserve: Major Threats and Management Plan. In *Bangladesh Environment 2000;* Feeroz Ahmed, M., Ed.; BAPA Bangladesh Poribesh Andolon: Bangladesh, 2000; pp 239–253.

FSP (Forestry Sector Project) 2000b; First Five Year Management Plan for Lawachara National Park, Background and Support Material, Forest Department, Ministry of Environment and Forests: Bangladesh, Vol. 2: 2000.

Integrated Protected Area Co-Management (IPAC) 2010; State of Bangladesh's Forest Protected Areas, 2010.

IUCN Bangladesh 2006; IUCN Red List of Threatened Species. 2006. (www.iucnredlist.org).

Khan, M. A. R. *On The Distribution of the Mammalian Fauna of Bangladesh,* Proceedings of the Second National Forestry Conference, Dhaka, Bangladesh, Jan 21–26, 1982; pp 560–575.

Khan, M. A. R. *Wildlife of Bangladesh;* Check List Cum Guide: Dhaka, Bangladesh, 2013.

Leech, J.; Ali, S. S. *Report of the Extended Natural Resources Survey: Part IV – Plant and Animal Species Lists. GoB/WB Forest Resources Management Project, Technical Assistance Component*; Mandala Agriculture Development Corporation: Dhaka, Bangladesh, 1997.

Mukul, S. A. *The Role of Traditional Forest Practices in Enhanced Conservation and Improved Livelihoods of Indigenous Communities: Case Study from Lawachara National Park, Bangladesh,* Proceedings of the 1st International Conference on Forest Related Traditional Knowledge and Culture in Asia, Seoul, Korea, Oct 5–10, 2008; pp 24–28.

Mukul, S. A. Biodiversity Conservation and Ecosystem Functions of Traditional Agroforestry Systems: Case Study from Three Tribal Communities in and Around Lawachara National Park. In *Forest Communities in Protected Areas of Bangladesh: Policy and Community Development Perspectives*;

Chowdhury, M. S. H., Ed.; Springer: Switzerland, 2014; pp 171–179.

Nishorgo Support Project (NSP), Bangladesh Forest Department. (http://www.nishorgo.org).

Rabbi, M. G.; Jaman, M. F.; Sarker, S. U. Ecology and Status of Avifauna of Nijhum Dwip and Damar Char, Noakhali and the Conservation Issues. *J. NOAMI*, **2011,** *28* (2), 59–71.

Thompson, P. M.; Johnson, D. L. Eds.; Checklist of Birds Recorded at 19 Sites in Bangladesh. Updated to Feb1, 1999. Unpublished Report. 1999.

CHAPTER 12

THE INFLUENCE OF ECOLOGICAL ATTRIBUTES AND MOTIVATIONS ON WILDLIFE TOURISM: THE CASE OF BIRDWATCHING IN BANDERAS BAY MEXICO

V. S. AVILA-FOUCAT*, A. SANCHEZ-VARGAS, and A. AGUILAR IBARRA

Instituto de Investigaciones Econymicas, Universidad Nacional Autynoma de Mexico, Mexico

**Corresponding author. E-mail: savila@iiec.unam.mx*

CONTENTS

ABSTRACT

Birdwatching is a specific sector of wildlife tourism that has increased in popularity around the world. The determinants of birdwatching demand in Mexico have not been analyzed, and to start this assessment the main objective of this article is to determine whether ecological attributes, motivations, past experience, and attitudes are relevant for birdwatching in Banderas Bay, Mexico. For that purpose, an on-site survey was carried out, and a count model was used to determine demand and an estimate of consumer surplus as a welfare measure of this economic activity. Results show that active birder category is more dominant than the casual or specialized categories but only environmental quality of the site and attitudes toward conservation is determinant for birdwatching demand. An estimate of visitors' economic welfare through their consumer surplus is of ~US$ 342–401. This study begins the process of understanding birdwatching demand determinants in Mexico showing that demand is determined by environmental attributes and attitudes and suggests some recommendations for the site in terms of market differentiation and complementary activities, as well as habitat and species conservation.

"Non-consumptive wildlife tourism demand depends on tourist typologies, site attributes, attitudes and socioeconomic aspects."

12.1 INTRODUCTION

Wildlife watching is a non-consumptive activity defined as "a human recreational engagement with wildlife wherein the focal organism is not purposefully removed or permanently affected by the engagement." Essentially, wildlife watching aims to increase the probability of positive encounters with wildlife while protecting wildlife resources (Reynolds & Braithwaite, 2001). The revenue provided by wildlife watching is relevant in some countries (Avila-Foucat et al., 2013), and Reynolds and Braithwaite (2001) have reported that ~40% of international tourists are wildlife related. Specifically, birdwatching is attractive to many tourists due to its few requirements in equipment and physical condition, which results in low costs and allows access to an increasing number of tourists (Eubanks et al., 2005). Hence, birdwatching has increased in popularity around the world (Maple et al., 2010). Wildlife tourism in México is concentrated on whales, birds, turtles, butterflies, and whale sharks. However, whale watching and whale shark observation are among the most important activities (Avila- Foucat et al., 2013). Because México has 11% of the world's bird species and 10% of them are endemic, it is an attractive destination for birdwatching, providing revenues of ~$23 million (Cantú & Sánchez, 2011).

Scholars have studied the motivations, specializations, and attitudes of birdwatching tourists as well as socioeconomic variables elsewhere (Catlin et al., 2011; Reynolds & Braithwaite, 2001; Weaver & Lawton, 2007; Curtin, 2013). However, there are limited socioeconomic studies regarding wildlife watching and birdwatching in Mexico. Particularly, the determinants of birdwatching demand in Mexico have not been assessed, and to start this assessment, a pilot survey was conducted for Banderas Bay, Mexico. The main objective of this chapter is to determine whether ecological attributes, motivations, past experience, and attitudes are relevant for birdwatching. For that purpose, a count model was used to determine not only the demand for visits but also an estimate of consumer surplus as a welfare measure of this economic activity.

12.2 BACKGROUND ON WILDLIFE TOURISM DETERMINANTS

The determinants of wildlife tourism have been widely studied. Reynolds and Braithwaite (2001) mention that successful wildlife tourism is based on tangible (infrastructure, services, and comfort) and intangible aspects from which the destination attributes, tourist's socioeconomic characteristics, specialization, motivations, and satisfaction are the most relevant (Catlin et al., 2011; Reynolds & Braithwaite, 2001; Weaver & Lawton, 2007; Curtin, 2013).

Nature site attributes have been studied by Deng et al. (2002), showing that natural resource conditions and accessibility are among the most important variables, followed by infrastructure, local community hospitality, and peripheral attractions. Wildlife watching demand depends not only on a site's ecological attributes but also on the quality of those attributes (Loomis, 1995; Guyer & Pollard, 1997; Deng et al., 2002; Huybers & Bennett, 2003). Huybers and Bennet (2003) analyzed in more detail the influence of environmental conditions (unspoiled, spoiled, and very spoiled), as well as crowding, prices, facilities, and the uniqueness of nature attractions in tropical North Queensland. They found that environmental conditions have a significant effect on tourists' demand.

Furthermore, Weaver and Lawton (2007) propose that ecotourism demand is based on tourist motivations and typologies resulting in market segmentation. There are several tourist typologies, from simple classifications such as casual versus serious (Eubanks et al., 2005) or hard versus soft core (Laarman & Durst, 1987) to a more extended classification including naturalistic, humanistic, moralistic, scientific, aesthetic, utilitarian, and dominionistic. Birder typologies such as casual, committed, and active are based on variables related to motivations, birding skills, past experience (frequency), expenses, and commitment to birdwatching (Scott et al., 2005; Cole & Scott, 1999; Martin, 1997).

For example, Scott et al. (1999) showed that searching for birds, being with friends, and wildlife conservation are the main motivations for highly skilled and committed birders. Additionally, Eubanks et al. (2004) found little variation in the socio-demographic characteristics of birders but found more variation in the type of tourists (committed, active, and causal viewer), based on their skills, motivations (e.g., improving birding skills, or watching birds that the viewer has not seen before), and expenditures. Moreover, Curtin (2013) indicates that for serious wildlife seekers, the primary motivation is wildlife (sometimes one species in particular), they preferring uniqueness, unspoiled countryside, and quiet places. In contrast, for casual birders, wildlife is not their primary motivation; nevertheless, they have interest in wildlife but lack of knowledge of what there is to see. Similarly, Cole and Scott (1999) showed that casual tourists expect less variety in flora and fauna and are more likely to participate in organized tour trips and complementary activities. Likewise, Scott and Thigpen (2003) demonstrated that birding involvement is related to the variety of birds, scenic beauty, other recreation activities, comfort amenities, and opportunities for nature hiking. Additionally, Martin (1997) showed that casual birders are more interested in wildlife watching infrastructure.

Leisure studies have shown that attitudes are relevant to predict satisfaction and future behavior (Hung-Lee, 2009). Thus, attitudes might be relevant birdwatching determinants. For instance, the theory of planned behavior (TPB) has been used to analyze the individual's intention to behave, to improve estimations of future or actual behavior (Han & Kim, 2010; Han et al., 2010). According to the TPB, the chief incentive to carry out any behavior is the intention to perform it, which depends on attitudes, subjective norms, and perceived behavioral control (Ajzen, 1991). For hunting, the intention to behave is related to attitudes, subjective norms (social pressure), and behavioral control (individual perception) (Hrubes et al., 2001). However, other studies have noted that attitudes differ among tourists and that the outdoor experience is optimized in different ways (Daigle et al., 2002). However, despite its potential role in explaining wildlife watching demand, the TPB has not been used to estimate either behavior or the intention to behave in birdwatching tourism.

In sum, non-consumptive wildlife tourism demand depends on tourist typologies, site attributes, attitudes, and socioeconomic aspects.

12.3 WILDLIFE WATCHING IN MEXICO

Wildlife watchers represent 36% of the nature tourists (wildlife watching and adventure tourism) in Mexico (CESTUR, 2006) and provide 26.5% of the revenues from this type of tourism. Although nature tourism may account for a low proportion of total tourism revenues, the federal government has established some programs to promote this activity, such as the National Program of Ecotourism and Rural Tourism created in 2010. The government has made such effort because wildlife tourism has enormous potential in this country due to Mexico's biodiversity and the significant tourism demand from the United States and Canada. Wildlife tourism in the United States has increased 13% from 1996 to 2006[1]. In fact, one out of every one hundred dollars of all goods and services produced in the United States is associated with wildlife recreation (USFWS, 2007). Moreover, 47% of wildlife observers are birdwatchers, and almost 40% of these observers travel away from home to view birds (USFWS, 2007). Moreover, in Canada, 30% of adults are wildlife observers, of which 7.5% are birdwatchers, and 13% of these had traveled to Mexico for wildlife observation in 2005 (Canadian Tourism Commission, 2006). International birdwatchers in Mexico generated economic benefits up to US$ 23 million in 2006 (Cantú & Sánchez, 2011). Therefore, birders represent an interesting market for wildlife tourism in Mexico. However, there are only a few studies on wildlife tourism in this country (Avila-Foucat & Pérez-Campuzano, 2015). The determinants of birdwatching tourism demand have not been studied, despite the high diversity of birds (11% of the

[1]Four out of every 10 people will participate in some type of wildlife recreation.

world's species), the existence of 43 birdwatching associations and 78,820 birdwatchers was identified in 2006 (Cantú & Sánchez, 2011).

12.4 METHODS

12.4.1 AREA OF STUDY

Banderas Bay is located on the Pacific coast of Mexico within the Jalisco and Nayarit states. Puerto Vallarta was the first pole of attraction for tourism, but the tourism infrastructure has now extended to the Nuevo Vallarta resort and Nayarit state. Local statistics show that the third most popular attraction enjoyed by tourists is the forest, after the beach and Puerto Vallarta's old town (Gobierno del Estado de Jalisco, 2007). Birding takes place in the semi-deciduous forest and in the Quelele lagoon. The lagoon has 100 hectares surrounded by white mangrove located 15 km from Puerto Vallarta. The bird diversity in the area includes ~366 species, of which 173 are resident birds and 148 are migrant birds, present during the winter. According to the national legislation, five bird species are in danger, and 12 species are endemic in this region (Gobierno del Estado de Nayarit, 2008).

Birdwatching is one of the non-consumptive wildlife activities in the area in addition to whale watching, snorkeling, and hiking. The main company providing birdwatching tours receives ~290 birders a year. National data show that wildlife watchers are mainly men from 25 to 34 years old, with a university education, with 70% usually staying in hotels and with an average annual income of US$ 40,000 (CESTUR, 2006). However, nature tourism is affected by the expansion of general tourism. Increasing tourism has brought economic benefits but also demographic growth and environmental deterioration (Márquez-González & Sánchez-Crispín, 2007; Gobierno del Estado de Nayarit, 2008). Proof of this deterioration is the 558 ha area that was deforested from 1980 to 2000, at a deforestation rate of 27.9 ha/y due to agriculture expansion, urbanization, and tourism infrastructure (Márquez-González, 2008).

12.4.2 SURVEY

An on-site structured pilot survey was applied to birdwatchers after their bird observation experience during the period from November 2007 to

March 2008 in Banderas Bay, Mexico. A total of 87 completed questionnaires were obtained, representing 30% of the birdwatchers in one year. Therefore, the data are representative of birdwatching in Banderas Bay. The survey included socioeconomic information, type of tourists, environmental attributes, and scenarios, as well as attitude variables. Because it was an on-site survey, it included structured questions. The variables are described in the following section.

12.4.3 SOCIOECONOMICS VARIABLES

General socioeconomic data, such as occupation, education, price of tours, income, as well as data on potential substitute tourism activities, were obtained. The travel costs for birdwatching were estimated based on the information on country of origin, type of accommodation, and length of stay in Banderas Bay.

12.4.4 MOTIVATION, PAST EXPERIENCE, AND TYPE OF TOURIST VARIABLES

As suggested by Eubanks et al. (2004), the tourists were divided in three types: committed, active, and casual viewers. In this chapter, the tourists were classified depending on their motivation in terms of their interest in nature and birding and on past birding experience. A committed birder is searching for contact with nature, has a specific interest in birding, has previous birding experience, and plans the birding activity before leaving home. An active birder has the same characteristics but is also interested in other complementary activities. A casual viewer is a person who does not plan birding but does it occasionally. Scott et al. (2005) described a variety of ways to define wildlife tourist typologies, indicating that there is no difference between self-classification and the other methods when identifying generic birdwatching motives (Scott et al., 2005). Therefore, in this study, tourist typologies are based on characteristics related to interest in nature, birding centrality, planning, and frequency (Table 12.1).

An interest in nature was assessed by asking birders to rank their reasons for travelling to Banderas Bay in order of importance (enjoying the beach and sun, enjoying hotel activities, having contact with nature, having contact with local communities, and adventure). Centrality was measured by asking the birders which other nature activities they planned to engage in during their trip. The planning and frequency of birding were evaluated by asking whether birdwatching was planned before leaving home and whether it was the first time birdwatching, and if not, how many times a year they practice it.

TABLE 12.1 Variables Used for Birdwatchers Typology.

	Committed	Active	Casual
Nature interest	Nature as travel motivation Only birdwatching activity	Nature as travel motivation Other nature activities during their trip	Nature not the main motivation Other nature activities
Birding planning	Planning before leaving home	Planning before leaving home	No planning before leaving home
Birding frequency	More than five times a year	More than five times a year	Less than five times a year

12.4.5 ENVIRONMENTAL VARIABLES

We used the bird population and deforestation rate as the main environmental attributes related to birdwatching. Bird species diversity and abundance were chosen as critical variables because they are the main targets of birdwatchers but also because they are indicators of habitat conservation (Meire & Dereu, 1990; Walker et al., 2009) and because they are linked to plant territories that are threatened. Finally, they occupy different levels in the food web (Padoa-Schioppa et al., 2006) and consequently, their absence would presumably change the ecosystem structure.

Clearing, fragmentation, and change in vegetation structure have a well-known influence on bird diversity, species richness, and community structure (Christiansen & Pitter, 1997; Padoa-Schioppa et al., 2006; Castelletta, et al., 2005; Scott, et al., 2006; Watson, et al., 2004; Martínez-Morales, 2005; Peh et al., 2005). Therefore, the local deforestation rate (Marquez-Gonzalez, 2008) and its relationship to the bird population as described by Martínez-Morales (2005) were used. We obtained an estimate of approximately four species per year lost due to a deforestation rate of 27.9 ha/y. The latter was used to construct two scenarios, the first showing only deforestation rates and the second displaying the effect of deforestation on bird populations.

1. The deforestation rate in the north of Banderas Bay is 27.9 ha/y, which is equivalent to 80% of the mangrove coverage in the Quelele lagoon, or ~27 professional football fields. If the deforestation rate continues (27 ha/y) will this affect your choice to return to Banderas Bay?
2. If the deforestation rate is 27 ha/y, the bird population will be affected and approximately four species a year will disappear. Will you be willing to return under such a scenario?

12.4.6 ATTITUDE VARIABLES

We included attitude variables according to the TPB (Ajzen, 1991), which, as noted above, has a potential role in explaining wildlife watching. Hence, attitude variables were designed according to the TPB to assess whether tourist behavior to repeat birdwatching is influenced by attitudes toward general conservation issues. Attitudes were thus measured by asking, "Do you think that environmental conservation should be a priority for humanity?" because it reflects a general environmental consciousness. Attitudes toward specific conservation actions were assessed with the question "Will you be interested in participating in natural resource conservation in Banderas Bay?" because it shows a specific and tangible intention toward conservation behavior compared with the previous question about environmental consciousness. Subjective norms were measured with the question "Is society worried about the environment in your country?" The aim here was to assess whether the visitor perceived a social pressure toward environmental conservation behavior. This is what Ajzen (Ajzen, 1991) defined as a "subjective norm." Finally, perceived behavioral control was assessed regarding the respondent's perceived role in conservation by asking "Do you consider that you have an important role in the environment conservation?" This variable "refers to people´s perception of the ease or difficulty of performing the behavior of interest" (Ajzen, 1991).

12.5 DEMAND MODEL

In this section, we briefly describe the count data approach for birdwatching demand estimation (Shaw, 1988; Grogger & Carson, 1991;

Cameron & Trivedi, 1998). This approach was used to offer reliable econometric evidence of wildlife watching demand in Banderas Bay. The objective of count models in the context of recreational demand is not only to determine the demand for visits but also to estimate a welfare measure such as the consumer surplus. The count data approach constitutes a good alternative to address the truncation problem often associated with samples drawn from on-site recreational surveys (e.g., Englin & Shonkwiler, 1995). On-site samples are often truncated and endogenously stratified because they do not include non-users and because the likelihood of certain persons being sampled depends on the frequency of their site visits, respectively. In this context, count data models, such as the Poisson model, allow us to estimate the demand for trips and, at the same time, to account for truncation and endogenous stratification. Moreover, it has been demonstrated that for recreational demand models, it is important to add not only the price of the recreation activity but also travel costs as a proxy of expenditure measurement (Loomis, 1995). The specification of our model is then as follows:

$$y_i^* = E(y_i^* \mid P_{i,j}, x_i, w_j, z_j) + e_i \tag{12.1}$$

Where y_i^* refers to the *i*th person's returning that is equivalent to a quantity demanded of the *j*th site, $P_{i,j}$ is the travel cost to the *j*th site[2], x_i and w_j are general individual and site characteristics, respectively, and z_j are attitude and environmental quality variables. Travel costs were included based on the expectation that a tourist will return depending not only on the price of the birdwatching tour but also on the costs to reach the site. Because y^* is the latent quantity demanded, we can obtain a consumer surplus estimate by integrating y^* over the relevant price change and environmental quality improvement (*dp*).

$$\int y^* \, dp \tag{12.2}$$

A more specific representation of the demand function (12.1) results from the following assumptions. First, given that on-site surveys collect

[2]Travel cost was estimated using local tourism statistics including flight prices, average expenses and tour price.

data from only a portion of the population (the people with non-zero demand) we observe only y_i as follows.

$$y_i = y_i^* \quad if \quad y_i^* > 0 \tag{12.3}$$

Moreover, if we assume that the conditional density $f(y_i^* \mid x_i)$ is Poisson with location parameter λ_i then the on-site sample's density function is given by

$$h(y_i / X_I) = \frac{e^{-\lambda_i} \lambda^{y_i - 1}}{(y_i - 1)!} \tag{12.4}$$

With the following conditional mean and variance

$$E(y \mid x_i) = \lambda_i + 1$$

$$Var(y \mid x_i) = \lambda_i$$

Here, the expected latent demand is estimated as a semi-logarithmic function of price, income, and other variables.

$$\ln \lambda_i = \beta_0 + \beta_p P + \beta_1 X_1 + ... + \beta_i X_i \tag{12.5}$$

To estimate the societal value of birding we estimated the consumer surplus. Hellersten and Mendelsohn (1993) showed that we can calculate the expected value of consumer surplus, *E(CS)*, derived from a count model as follows:

$$E(CS) = \frac{E(y_i / x_i)}{-\beta_P} = \frac{\hat{\lambda}_i}{-\beta_P} \tag{12.6}$$

where $\hat{\lambda}_i$ is the expected number of trips and β_P is the sensitivity of the response variable (i.e., *demand*) to price changes. Thus, the per-trip consumer surplus is equal to

$$\frac{1}{-\beta_P}.$$

Finally, to estimate the demand and the associated consumer surplus, we use maximum likelihood methods, based on equation (12.3). A criticism of the Poisson model is that the mean–variance equality, conditional on explanatory variables is often violated with over-dispersion of the dependent variable and a model with negative binomial distribution would provide a better fit if that is the case. Therefore, a test of the validity of the mean and variance must be reported to ensure that the Poisson model is an appropriate specification for the data at hand. Thus, for the econometric analysis, we specified the demand (λ_i) for trips to Banderas Bay as:

$$\ln \lambda_i = \beta_0 + \beta_p P + \beta_1 X_1 + \beta_2 X_2$$

where λ_i is the number of trips taken by individual i, P denotes travel costs, X_1 is the individual's household income, and X_2 is a vector that includes variables such as individual, TPB, and place characteristics including environmental quality.

12.6 RESULTS

The interviewee's general information shows that birdwatchers are mostly international tourists from Canada (46%) and the United States (50%). Their average age is 56 years old, 43% of them have postgraduate studies, and 41% have an income of more than US$ 72,000 a year. Similar results for birdwatchers in Mexico are mentioned by Cantú and Sánchez (2011). In our sample, 69% were women and the majority of tourists were lodged in hotels in Puerto Vallarta (67%) or renting a condominium.

Our results show that birdwatchers are mostly either committed birders or active birders because they plan birding before leaving home, nature is their main motivation for travelling, and the birding frequency is more than five times a year for 58% of the sample. Nature is extremely important for 27% and moderately important for 35% of the birdwatchers, and 50% of the birdwatchers performed nature activities other than birding such as whale watching or rafting. Thus, active birders are dominant over committed or casual birders following our classification (Table 12.2).

The estimated parameters for the Poisson count model for assessing the determinants of birdwatching demand are presented in Table 12.3, where we show the significance of socioeconomic variables, environmental attributes, and attitudes. The type of tourists was not a significant variable.

TABLE 12.2 Results of Birdwatchers Typology.

Variables/type of tourists	Committed	Active	Casual
Nature interest	Nature is extremely important (27%)	Nature is moderately important (35%)	Nature is not the main motivation (42%)
Activities during their trip	Only birdwatching (8%)	Other nature activities during their trip (50%)	Other nature activities during their trip (50%)
Birding planning	Planning before leaving home (65%)		No planning before leaving home (45%)
Birding frequency	More than five times a year (58%)		Less than five times a year (42%)

TABLE 12.3 Poisson Model Results for Describing Determinants of Birdwatching Demand in Banderas, Bay. Significant Coefficients ($p < 0.1$) are Presented and Standard Errors are Shown in Parentheses.

Variables	Poisson	Truncated Poisson
Socioeconomic variables		
Age	0.0369 (0.0062)	0.0538 (0.0157)
Price of travel	−0.00017 (0.00005)	−0.0002 (0.0001)
Income	−0.0001 (0.00006)	−0.0002 (0.0001)
Environmental variables		
Deforestation increases and bird species decrease	−0.3193 (0.1742)	−0.6267 (0.3666)
Number of birds	0.1329 (0.0681)	0.2619 (0.0972)
Attitude variables		
Participating in conservation	0.1475 (0.0835)	0.3520 (0.1978)
Over-dispersion test	0.9605	–
Log pseudo-likelihood	−28.84229	−48.5454

The model shows that age is a significant and positive variable, confirming that demand increases when birdwatchers are older. The income parameter ("Income") presents a negative coefficient but close to a zero value, suggesting that income does not have an important influence on the decision to return. The negative coefficient of the price variable ("Price") implies that as the travel cost increases, demand decreases.

Specialization does not determine the demand for birdwatching in our sample because nature interest, centrality, or frequencies are not significant. In contrast, demand for trips is rather sensitive to variables associated with ecological attributes, deforestation rates, and individual characteristics. The response of demand to changes in the number of birds is positive and higher than the response to prices and income. Moreover, it is interesting to note that the number of birds is more relevant than other ecological attributes such as habitat, plant diversity, bird species, or the number of endemic species. Another interesting result is the significance of deforestation and a scenario of decreasing bird species (coefficients: −0.32 and –0.63 in the Poisson and truncated Poisson models, respectively), which implies that under this scenario, the number of visitors decreases, confirming our hypothesis that the deterioration of environmental assets plays a very important role in the demand for wildlife observation.

The model shows that only attitude toward participating in conservation activities was significant to birdwatching demand. This finding is interesting because the TPB has proven to be a significant indicator of the intention to behave (Chan, 1998). According to other studies, attitude is the most significant variable with respect to subjective norms and perceived behavioral control (Karpinnen, 2005; Chan, 1998; Fielding et al., 2008, Pouta & Rekola, 2001). Further, the quantity of trips demanded by users presented in Table 12.4 is almost two per year, indicating that birdwatchers repeat visits to Banderas Bay. This number is estimated using the sample means of the variables. Moreover, the expected consumer surplus estimates (US$ 401 according to the Poisson model and US$ 342 according to the truncated Poisson model) indicate that tourists would be willing to pay more for birdwatching in Banderas Bay. Such amounts justify (other things being equal) the intention to visits almost twice a year, as shown by the model results.

TABLE 12.4 Estimated Quantity Demanded of Trips and Consumer Surplus.

	Poisson	Truncated Poisson
User's quantity demanded (number of trips per person per year)	1.657	1.747
Consumer surplus (USD)	$401.0	$341.6

12.7 DISCUSSION

It is surprising that the type of tourist or specialization does not determine birdwatching demand in our study and neither does nature interest, birding frequency, or planning. Nevertheless, birders are repeating their visits for birding or for engaging in other activities. Indeed, 50% of our interviewees were whale watching, snorkeling, and hiking as complementary activities. Scott and Thirpe (2003), Hvenegaard (2002), Curtin (2013), Martin (1997), and Cole and Scott (1999) have shown that complementary activities are important for birders and vary with specialization. Furthermore, planning birding before leaving home is not significant, meaning that birders are willing to return independently of their specialization, probably due to the experience they had birdwatching. These results differ from those of other studies, which shows a difference in birders' motivations depending on specialization level (Scott et al. 2005; Hvenegaard, 2002; Scott et al., 1999; Scott & Thigpen, 2003; Martin, 1997; Cole & Scott, 1999) and shows that the market is segmented for different types of consumers. In Banderas Bay, the results show that even if there are different types of birders, they are all willing to return. This suggests that a more detailed classification is needed, but there is also an opportunity to differentiate markets based on the type of tourist.

In Banderas Bay, socioeconomic variables and site attributes are comparatively more relevant. Environmental attributes, measured by deforestation level and the decrease in bird species are negatively related to demand, showing that birdwatchers are interested in both attributes (i.e., birds and their habitat), as might be expected due to their commitment to birds and conservation. This is in accordance with Hvenegaard (2002), who showed that specialization level is positively related to conservation involvement. In addition, the model results show that the number of birds is more significant than species rarity or diversity, suggesting that birders are not looking for species specific to Banderas Bay, thus increasing the probability of substituting another birdwatching destination for Banderas Bay. Although the type of tourist is not significant in the model, having active viewers instead of committed viewers would explain why birders are more interested in the number of birds rather than in the species. Martin (1997) obtained similar results, showing that intermediate birders prefer to see wildlife closely, with rare and different species, compared to novice and specialists. However, our results are in contrast to those of Catlin et al. (2011), who argue that tourists are more attracted to rare or uncommon species. Similarly, Scott and Thigpen (2003) show that active birder are more interested in bird diversity than casual or skilled birders. Cole and Scott (2008) demonstrate that serious birders are more interested in the variety of species and native birds.

Identifying the relationship between environmental attributes of the site and the type of tourist is not an easy task. The context of each study needs to be considered when analyzing and comparing case studies because an active birder could be a serious birder in another context even using the same classification or typology. For instance, Martin (1999) applied a survey for classifying tourists as Montana non-residents. In contrast, Cole and Scott (2008) considered a casual birder to be anyone with a conservation passport and a serious birder to be any member of the American Birding Association. The negative sign of income and travel costs in the model shows that Banderas Bay birdwatching demand fluctuates compared to that of other birding sites.

Attitude toward conservation is positively related to birdwatching demand in accordance with previous studies. Attitude has proven to be the most significant variable with respect to subjective norms and perceived behavioral control (Karpinnen, 2005; Chan, 1998; Fielding et al., 2008; Pouta & Rekola, 2001). That is, a visitor who thinks that conservation is important returns to birding in Banderas Bay. This can be explained based on Hvenegaard's (2002) results showing that a specialized birder has more conservation involvement (e.g., donating to a conservation association). In contrast, social pressure (subjective norms) and perceived behavioral control are not significant probably because our sample is mainly composed of active and committed viewers, for whom the decision to do birding is independent from the social consciousness in their country or their role in conservation. This is in contrast to the findings of Avila-Foucat et al. (2013), that a subjective norm was a significant variable in returning to whale watching in Banderas Bay. Thus, the relationship between non-consumptive wildlife demand and the type of tourists and their attitudes needs to be explored further. Finally, the birdwatcher's consumer surplus in the truncated model is US$ 341.60, which is an amount similar to the price of an airplane ticket from Canada to Banderas Bay. Similar evidence in Texas demonstrates that active birding tourists have a higher willingness to pay than casual tourists, as shown by Eubanks et al. (2004).

Duffus and Dearden (1990) explain that wildlife tourism destinations attract specialists first and that destinations are then progressively dominated by generalists, who require more facilities and mediation between themselves and the target species. Therefore, birdwatching in Banderas Bay might be in the beginning stages and eventually can evolve in a sustainable way, if over-crowding and environmental impacts are minimized. For instance, the whale watching in our study area is an industry with fewer specialized wildlife tourists that has suffered from congestion, which has had a significant influence on potential demand (Avila-Foucat et al., 2013).

The limitations of this study are first related to the scale since the study that it is not representative for the birdwatching activity in all Mexico, but a similar study could be repeated in different birding locations. Another limitation is that birder typology characteristics were predetermined and a self-specialization could be explored in further studies including other variables such as birdwatching skills. The characteristics of specific birds can also be explored, such as approachability, predictability, and diurnal activity, as suggested by Reynolds and Braithwaite (2001).

However, this study begins the process of understanding birdwatching demand determinants and suggests some recommendations for the site. The first important finding for managers is that the active birder category is more dominant than the casual or specialized categories. Moreover, because birders are interested in birdwatching combined with complementary activities, the commercialization of birding can be linked to other nature activities. The second important finding is that the market can be differentiated to target the different types of birders and that different strategies can be created to promote birdwatching. Additionally, demand is determined by environmental attributes and attitudes; thus, if local government wants to promote sustainable tourism, conservation of species and their habitat is required. It might be possible to create a local participatory program for conservation action because attitudes toward conservation are positively related to demand. Consumer surplus shows that birdwatchers are willing to pay more for this activity and that it is a valuable asset.

12.8 CONCLUSION

The environmental quality of the site and attitudes toward conservation are determinants for birdwatching, a non-consumptive wildlife tourism activity, in Banderas Bay, Mexico. Birdwatchers are mainly active viewers, and this is important data for market segmentation. However, in this model, specialization is not significant for birdwatching demand. We also provided an estimate of visitors' economic welfare through their consumer surplus of ~US$ 342–401. This is, in fact, a way of valuing birdwatching in Banderas Bay. A positive consumer surplus indicates that tourists are willing to pay more than they currently do. Hence, tourists repeat visits, at least once a year, indicating that birdwatching is a valuable asset for wildlife tourism in this region. Finally, the hypothesis that the demand for bird observation depends not only on economic variables but also on attitudes and environmental quality assets is an interesting finding for the case of Banderas Bay in Mexico.

KEYWORDS

- **birdwatching**
- **tourists' preferences**
- **demand**
- **Mexico**
- **wildlife watching**

REFERENCES

Ajzen, I. The Theory of Planned Behavior. *Organ. Behav. Hum. Decis. Process.* **1991,** *50,* 179–211.

Avila -Foucat, V. S.; Sánchez Vargas, A.; Frisch Jordan, A.; Ramírez Flores, O. M. The Impact of Vessel Crowding on the Probability of Tourists Returning to Whale Watching in Banderas Bay, Mexico. *Ocean Coastal Manag.* **2013,** *78,* 12–17. doi: j.ocecoaman.2013.03.002

Avila-Foucat, V. S.; Perez-Campuzano, E. Municipality Socioeconomic Characteristics and the Probability of Occurrence of Wildlife Management Units in Mexico. *Environ. Sci. Policy.* **2015,** *45,* 146–153. doi: j.envsci.2014.08.005

Cameron, A. C.; Trivedi, P. K. Regression Analysis of Count Data. *Econ. Soc. Monogr.* **1998,** *30,* 432.

Canadian Tourism Commission 2006; Travel Activities and Motivation Survey: Canadian Activity.

Profile, Wildlife Viewing While on Trips, Canada, 2006.
Castelletta, M.; Thiollay, J. M.; Sodhi, N. S. The Effects of Extreme Forest Fragmentation on the Bird Community of Singapore Island. *Biol. Cons.* **2005,** *121,* 135–155.
Catlin, J.; Jones R.; Jones, T. Revisiting Duffus and Dearden´s Wildlife Tourism Framework. *Biol. Cons.* **2011,** *144,* 1537–1544.
Cantú, J. C.; Sánchez, M. E. Observación de Aves: Industria Millonaria. *Biodiversitas.* **2011,** *97,* 20–15.
Centro de Estudios Superiores de Turismo (CESTUR) 2006; Perfil y Grado de Satisfacción del Turista Que Viaja en México por Motivos de Ecoturismo, CESTUR: México, 2006.
Chan, K. Mass Communication and Pro-Environmental Behavior: Waste Recycling in Hong Kong. *J. Environ. Manag.* **1998,** *52,* 317–325.
Christiansen, M. B.; Pitter, E. Species Loss in a Forest Bird Community Near Lagoa Santa in Southeastern Brazil. *Biol. Cons.* **1997,** *89,* 23–32.
Curtin, S. Lessons from Scotland: British Wildlife Tourism Demand, Product Development and Destination Management. *JDMM.* **2013,** *2,* 196–211.
Cole, J. S.; Scott, D. Segmenting Participation in Wildlife Watching: A Comparison of Casual Wildlife Watchers and Serious Briders. *Hum. Dimens. Wildl.* **1999,** *4* (4), 44–61.
Daigle, J.; Hrubes, D.; Ajzen, I. A Comparative Study of Beliefs, Attitudes, and Values among Hunters, Wildlife Viewers and other Outdoor Recreationists. *Hum. Dimens. Wildl.* **2002,** *7,* 1–19.
Deng, J. Y.; King, B.; Bauer, T. Evaluating Natural Attractions for Tourism. *Ann.Tour. Res.* **2002,** *29* (2), 422–438.
Duffus, D. A.; Dearden, P. Non-Consumptive Wildlife Oriented Recreation: A Conceptual Framework. *Biol. Cons.* **1990,** *53,* 213–231.
Englin, J.; Shonkwiler, J. S. Estimating Social Welfare Using Count Data Models: An Application to Long-Run Recreation Demand Under Conditions of Endogenous Stratification and Truncation. *Rev. Econ. Stat.* **1995,** *77* (1), 104–112.
Eubanks, T. L.; Stoll, J. R.; Ditton, R. B. Understanding the Diversity of Eight Birder Sub-Populations:Socio-Demographic Charactersitics, Motivations, Expenditures and Net Benefits. *J. Ecotour.* **2005,** *3* (3), 151–171.
Fielding, K. S.; McDonald, R.; Louis, W. R. Theory of Planned Behavior, Identity and Intentions to Engage in Environment Activism. *J. Environ. Psychol.* **2008,** *28,* 318–326.
Gobierno del Estado de Jalisco 2007; Anuario Estadístico del Estado de Jalisco: Jalisco, Mexico, 2007.
Gobierno del Estado de Nayarit 2008; Plan Municipal de Desarrollo de Bahía de Banderas 2005–2008: Nayarit, Mexico, 2008.
Grogger, J. T.; Carson, R. T. Models for Truncated Counts. *J. Appl. Econom.* **1991,** *6* (3), 225–238.
Guyer, C.; Pollard, J. Cruise Visitor Impressions of the Environment of the Shannon-Erne Waterways System. *J. Environ. Manag.***1997,** *51* (2), 199–215.
Han, H.; Kim, Y. An Investigation of Green Hotel Customers' Decision Formation: Developing an Extended Model of the Theory of Planned Behavior. *Int. J. Hosp. Manag.* **2010,** *29,* 659–668.
Han, H.; Hsu, L. T.; Sheu, C. Application of the Theory of Planned Behavior to Green Hotel Choice: Testing the Effect of Environmental Friendly Activities. *Tour. Manag.* **2010,** *31,* 325–334.

Hellersten, D; Mendelsohn, R. A Theoretical Foundation for Count Data Models, with an Application to a Travel Cost Model. *Am. J. Agr. Econo.* **1993,** *75,* 604–611.

Huybers, T.; Bennett, J. Environmental Management and the Competitiveness of Nature-Based Tourism Destinations. *Environ. Resour. Econ.* **2003,** *24,* 213–233.

Hung-Lee, T. A Structural Model to Examine how Destination Image, Attitude, and Motivation Affect

the Future Behavior of Tourists. *Leisure Sci.* **2009,** *31* (3), 215–236.

Hrubes, D.; Ajzen, I.; Daigle, J. Predicting Hunting Intentions and Behavior: An Application of the Theory of Planned Behavior. *Leisure Sci.* **2001,** *23,* 165–178.

Hvenegaard, G. T. Birder Specialization Differences in Conservation Involvement, Demographics and

Motivations. *Hum. Dimens. Wildl.* **2002,** *7,* 21–36.

Karpinnen, H. Forest Owners′ Choice of Reforestation Method: An Application of the Theory of Planned Behavior. *Forest Policy Econ.* **2005,** *7,* 393–409.

Laarman, J. G.; Durst, P. B. *Nature Travel and Tropical Forest;* FPEI Working Paper Series, Southeastern Centre for Economic Research: North Carolina State University, Raleigh, NC, 1987.

Loomis, J. B. Four Models for Determining Environmental Quality Effects on Recreational Demand and Regional Economics. *Ecol. Econ.* **1995,** *12,* 55–65.

Maple, L. C.; Eagles, P. F. J.; Rolfe, H. Birdwatchers' Specialization Characteristics and National Park. *J. Ecotour.* **2010,** *9* (3), 219–232.

Marquez- Gonzalez, A. Cambio de uso del Suelo y Desarrollo Turístico en Bahía de Banderas Nayarit. *Ciencia.* **2008,** *XI* (2), 161–168.

Márquez-González, A. R.; Sánchez-Crispín, A. Turismo y Ambiente: La Percepción de los Turistas Nacionales en Bahía de Banderas, Nayarit, México. *Boletín del Instituto de Geografía, UNAM.* **2007,** *64,* 134–152.

Martin, S. R. Specialization and Differences in Setting Preferences among Wildlife Viewers. *Hum. Dimens. Wildl.* **1997,** *2* (1), 1–18.

Martínez-Morales, M. A. Landscape Patterns Influencing Bird Assemblages in a Fragmented Neotropical Cloud Forest. *Biol. Cons.* **2005,** *121,* 117–126.

Meire, P. M.; Dereu, J. Use of the Abundance/Biomass Comparison Method for Detecting Environmental Stress: Some Considerations Based on Intertidal Macrozoobenthos and Bird Communities. *J. Appl. Ecol.* **1990,** *27,* 210–223.

Padoa-Schioppa, E.; Baietto, M.; Massa, R.; Bottoni, L. Bird Communities as Bioindicators: The Focal Species Concept in Agricultural Landscape. *Ecol. Indic.* **2006,** *6,* 83–93.

Peh, K. S. H.; de Jong, J.; Sodhi, N. S.; Lim, S. L-H.; Yap, A-M. Lowland Rainforest Avifauna and Human Disturbance: Persistence of Primary Forest Birds in Selectively Logged Forests and Mixed-Rural Habitats of Southern Peninsular Malaysia. *Biol. Cons.* **2005,** *123,* 489–505.

Pouta, E.; Rekola, M. The Theory of Planned Behavior in Predicting Willingness to Pay for Abatement of Forest Regeneration. *Soc. Natur.Resour.* **2001,** *14* (2), 93–106.

Reynolds, P. C.; Braithwaite, D. Towards a Conceptual Framework for Wildlife Tourism. *Tour. Manag.* **2001,** *22,* 31–42.

Scott, D. M.; Brown, D.; Mahood, S.; Denton, D.; Silburn, A.; Rakotondraparany, F. The Impacts of Forest Clearance on Lizard, Small Mammals and Bird Communities in the Arid Spiny Forest, Southern Madagascar. *Biol. Cons.* **2006,** *127,* 72–87.

Scott, D.,;Ditton, R. B.; Stoll, J. R.; Eubanks, T. L. Measuring Specialization among Birders: Utility of a Self-Classification Measure. *Hum. Dimens. Wildl.* **2005,** *10,* 53–74.

Scott, D.; Menzel Baker, S.; Kim, C. Motivations and Commitments among Participants in the Great Texas Birding Classic. *Hum. Dimens. Wildl.* **1999,** *4* (1), 50–67.

Scott, D.; Thigpen, J. Understanding the Birder as Tourists: Segmenting Visitors to the Texas Hummer/Bird Celebration. *Hum. Dimens. Wildl.* **2003,** *8,*199–218.

Shaw, D. On-Site Sample Regression: Problems of Non-Negative Integers, Truncation, and Endogenous Stratification. *J. Econom.***1988,** *37,* 211–223.

United States Fish and Wildlife Service (USFW) 2007; National Survey of Fishing, Hunting, and Wildlife-Associated Recreation, National Overview, United States Fish and Wildlife Service: US, 2007.

Walker, T. R.; Crittenden, P. D.; Dauvalter, V. A.; Jones, V.; Kuhry, P.; Loskutova, O.; Mikkola, K.; Nikula, A.; Patova, E.; Ponomarev, V. I.; Pystina, T.; Rätti, O.; Solovieva, N.; Stenina, A.; Virtanen, T.; Young, S. D. Multiple Indicators of Human Impacts on the Enviornment in the Pechora Basin, North-Eastern European Rusia. *Ecol. Indic.* **2009,** *9,* 765–779.

Watson, J. E. M.; Whittaker, R. J.; Dawson, T. P. Habitat Structure and Proximity to Forest Edge Affect the Abundance and Distribution of Forest-Dependent Birds in Tropical Coastal Forests of Southeastern Madagascar. *Biol. Cons.* **2004,** *120,* 311–327.

Weaver, D. B.; Lawton, L. J. Twenty Years on: The State of Contemporary Ecotourism Research. *Tour. Manag.* **2007,** *28,* 1168–1179.

CHAPTER 13

ECONOMIC VALUATION OF WILDLIFE TOURISM: "CONTINGENT VALUATION METHOD"

R. M. WASANTHA RATHNAYAKE*

Department of Tourism Management, Sabaragamuwa University of Sri Lanka, Belihuloya, Sri Lanka

**E-mail: warath1@gmail.com*

CONTENTS

ABSTRACT

This chapter describes how the contingent valuation approach can be applied to the economic valuation of wildlife tourism. The first part of the chapter describes the contingent valuation method (CVM), and reviews the literature on empirical studies of wildlife management and wildlife tourism from around the world. In the second part of the chapter a case study from Sri Lanka of wildlife tourism is discussed. The third part of the chapter discusses the issues and biases related to CVM and the National Oceanic and Atmospheric Administration (NOAA) guidelines in order to ensure the reliability and usefulness of the information obtained from CVM applied studies.

> **"...income is an important factor affecting the WTP for visitor facilities and/or turtle conservation."**

13.1 INTRODUCTION TO CONTINGENT VALUATION METHOD (CVM)

Welfare changes resulting from quality changes in recreational experiences are frequently estimated through revealed preference and stated preference valuation methods (Whitehead et al., 2000). CVM is a stated preference direct technique; the individual "states" his or her preference. The word "contingent" in the expression CVM means hypothetical.

By CVM we mean that the value of an environmental good is elicited directly through an answer to a question about a consumer's willingness to pay (WTP) to have more of a good, or willingness to accept (WTA) having less of it. Over the last 30 years, CVM has become the most used valuation method and has been developed mainly in the context of environmental valuation. There is therefore great interest in studying it more carefully.

CVM is often employed because it enables the estimation of a (Hicksian) measure of consumer surplus (CS) under circumstances when an environmental quality change is hypothesized or being planned (Hanemann, 1984; Mitchell & Carson, 1989; Hanemann et al., 1991; Cummings & Harrison, 1994). Specifically, with CVM individuals are asked about the status quo versus some alternative scenario. Information is elicited about how the individual "feels" about the alternative relative to the status quo, and their WTP, if anything, for the alternative. Contingent valuation (CV) is a method of recovering information about preferences or WTP from direct questions. The purpose of CV is to estimate individual WTP/WTA for changes in the quantity or quality of goods or services, as well as the effect of covariates on WTP.

13.2 STEPS IN CVM

The CVM is a survey-based tool. The survey instrument seeks to create a hypothetical market and the response to this market allows researchers to estimate respondents' values associated with an environmental good. There are five steps in establishing CVM.

13.2.1 STEP 1: CONSTRUCTION OF A HYPOTHETICAL MARKET

The main idea here is to construct a scenario ("hypothetical market") that corresponds as closely as possible to a real-world situation. It is still usually hypothetical. In most cases there will not be a direct link between the answers given by interviewees and any potential decision to implement or not the environmental change to be valued.

a) **Set the reason for payment:** We must pay to get more of a good. The improvement specified is contingent on payment actually being made. This scenario must be understood by respondents.
b) **Construct a so called bid vehicle or method of payment:** This vehicle must fulfill conditions with respect to incentive compatibility, realism, and subjective justice among respondents. Relevant vehicles are: a direct sum of money to be paid; payment to a fund/contribution; support of a particular tax and payment in the form of a higher price for a commodity related to the improvement (such as, higher electricity prices when the objective is to divert a river or nuclear power plant from being developed).
c) **Construct a provision rule:** This is a mechanism for the provision of the good, as a function of the stated value.

13.2.2 STEP 2: OBTAINING DATA

We select a limited sample of the underlying population, and take this sample through an interview (or possibly a sequence of interview sections). Interviews can be carried out: person-to-person interviews; using an interactive medium (computer); mail questionnaire (with follow-ups); or telephone interview. Most recommendations for research methodology begin with person-to-person interviews, which have the advantage of: face-to-face contact; increased engagement and awareness by the interviewee; reduced misunderstanding; and the opportunity to ask spontaneous questions. In some cases a computer program may be better at choosing a (complex) path of questions when there are several alternatives.

Data are collected to estimate both (a) the maximum WTP for an improvement in environmental quality or minimum WTA to abstain from an improvement in environmental quality and (b) WTP to avoid a worsening in environmental quality or WTA to accept a worsening in environmental quality. Most studies adopt WTP as the preferred valuation measure, due to perceived severe problems with WTA (protest bids and infinitely-high bids).

WTA questions are, however, often problematic, as there can be lot of emotion involved in answering them. It has been shown that a good may be valued quite differently, according to whether the individual initially does not have it (and must pay to get it), or whether the individual already has the good initially (and must give it up, in fact be "bribed" into giving it up). In the latter case, we often find very high valuation statements.

13.2.3 POSSIBLE BIDDING MECHANISMS

The final element of a CV scenario is the method of asking questions. This part of the questionnaire presents the respondent with a given monetary amount, and his or her response is sought. The literature on CV is replete with terms for survey practice and econometrics. The basic approaches to asking questions that lead directly to WTP or provide information to estimate preference follow.

1. **Open ended CV:** A CV question in which the respondent is asked to provide the interviewer with a point estimate of his or her WTP. "What is the maximum amount you would pay as an entrance fee

to enter the Rekawa Sanctuary under the Scenario 1?" This type of question is called "open-ended" CVM, "Open" because the respondent is permitted to say any amount that they want.

2. **Bidding game:** A CV question format in which individuals are iteratively asked whether they would be willing to pay a certain amount. The amounts are raised (lowered) depending on whether the respondents were (were not) willing to pay the previously offered amount. The bidding stops when the iterations have converged to a point estimate of WTP.
3. **Payment card:** A CV question format in which individuals are asked to choose a WTP point estimate (or a range of estimates) from a list of values predetermined by the surveyors, and shown to the respondents on a card. "On the following payment card (a card with a range of different USD amounts) circle the largest amount you would be willing to pay as an entrance fee to enter the Rekawa Sanctuary."
4. **Close ended questions**: There are at least three variants.
   - **Dichotomous choice (referendum):** A single amount is offered and respondents are asked to provide "yes" or "no" answer, also referred to as the take it or leave it approach.
   - **Double bounded referendum:** Respondents who answer "no" to the amount are offered a lower amount and those who answer "yes" are offered a higher amount.
   - **Trichotomous choice:** Respondents are offered three choices to the payment—"yes," "no," and "indifferent."

Out of these, "Dichotomous choice (referendum)"is usually considered incentive compatible and free of starting point bias, but provides little information (only one bound). The others provide more information, but the information may be distorted.

13.2.4 STEP 3: ESTIMATING AVERAGE WTP/WTA

Estimating is straightforward with open ended and bidding game formats. There are more difficulties with the single-bounded referendum, because to estimate probability functions more data are required.

13.2.5 STEP 4: ESTIMATING BID CURVES

Define the bid curve for individual i as follows:

$$WTP\,(i) = \int (hhince(i),\, educ(i),\, age(i),\, xs(i),\, quality(i)\ldots..)$$

where $hhince$ = household income, $educ$ = education, age = age, xs = vector of other background variables we want to include, and $quality$ = environmental quality.

The objective is to find a "best" fitting function of this sort, from the material collected. Since material is "experimental," simple estimation methods are usually sufficient (ordinary least square [OLS] with direct bid data, logit, or probit with referendum type data).

13.2.6 STEP 5: AGGREGATING THE DATA

To aggregate the data convert mean bids to population aggregates and use derived bids and bid functions for benefit transfer.

13.3 EMPIRICAL EVIDENCE OF THE APPLICATION OF CVM IN WILDLIFE TOURISM

Scholars have widely applied CVM to identify an entrance fee for a particular recreational service in a protected area. In order to assess WTP in a hypothetical wildlife tourism scenario, we need to describe carefully the conditions of the market and what is to be valued. In this instance, our interest is in the change in fees that tourists may be willing to pay if there is a change in the recreational service provided. As available research shows (Clawson & Knetsch, 1966; Leuschner et al., 1987; Reiling et al., 1988; Laarman & Gregersen, 1996; Lee, 1997; Vogt & Williams, 1999; Williams et al., 1999; Eagles et al., 2002), giving information to visitors on why a fee must be levied, and how the money will be utilized, is likely to affect positively their support for the fee-paying option and increase demand for the recreational service.

The literature underscores the demographic and psychographic factors that affect people's demand for changes in recreational services. Variables such as income, age, attitude toward environmental protection, history of paying entrance fees, and education are predicted to influence WTP (Reiling et al., 1992; Bowker et al., 1999; Williams et al., 1999; More & Stevens, 2000). According to Davis and Tisdell (1998) the effects of gender, country of residence and previous visits to the site, and to natural attractions in general, could be either positive or negative, requiring further testing. Moreover, as Schroeder and Louviere (1999) recognize, people are likely to pay more to enter a site if they have travelled a long distance to reach it.

Mitchell and Carson (1989) have identified a behavioral dimension to WTP, which they call behavioral intention. The findings of Kerr and Manfredo (1991) from a study of backcountry hut users in New Zealand's parks suggest that previous fee-paying behavior affects paying intention. In addition, membership in environmental organizations and attitude toward environment protection may also affect payment intention (Carlsson & Johansson-Stenman, 2000; Clinch & Murphy, 2001). As Laarman and Gregersen (1996) point out, what consumers expect to pay is related to what they have paid previously.

13.4 A CVM APPLICATION: A SRI LANKAN CASE STUDY

The application of CVM in wildlife tourism or wildlife conservation is typified by Rathnayake's (2015) *turtle study* conducted at Rekawa sanctuary in Sri Lanka.

13.4.1 BACKGROUND TO THE STUDY

The Rekawa beach is ~4 km long and is a prime turtle-nesting habitat in Sri Lanka. It has therefore been declared as a sanctuary under the Fauna and Flora Protection Ordinance (1937). In Rekawa, local Sri Lankans are dependent on turtles for their livelihood and therefore turtle conservation is difficult. For some low-income families in the fishing community turtle meat, shells, and eggs constitute their only source of income, with members of the community either consuming or selling turtle meat and eggs (IUCN, 2005). The Department of Wildlife Conservation in Sri Lanka has declared this beach a sanctuary in order to protect the breeding grounds. In addition, non-governmental organizations such as the "Turtle Conservation Project" (TCP) have started *in situ* conservation programs in the area. Furthermore, tourism related to turtle watching is growing in Rekawa. This study estimates the entrance fee that can be charged to visitors for "turtle watching" to ascertain whether revenues from such fees can be used to compensate the fishing community and to reduce illegal activities. This study employed CVM to examine changes in demand (or tourists WTP) and resultant changes in welfare as a result of a proposed change in the management of the Rekawa sanctuary.

13.4.2 QUESTIONNAIRE AND SCENARIO DEVELOPMENT

The survey instrument was a questionnaire, which sought to create a hypothetical market. The response to this market allows researchers to estimate respondent's values associated with an environmental good. The survey instrument, which is crucial in CVM, generally has three components: (a) a description of the choice setting (scenario/hypothetical situation), (b) the question from which values will be inferred, and (c) demographic questions about the respondents. The components (a) and (c) are common for all instruments.

The essential prerequisite of CV analysis is the design of the questionnaire and survey procedure. A CV question asks a respondent for a monetary valuation of a service that is meaningful to the respondent. The service must be limited geographically and temporally, and be defined in terms of characteristics that can reasonably enter a respondent's preference function. For example, to study WTP for a recreational service in a protected area or conservation program the question should be worded in a way commonly understood by the general public.

The scenario will attract more visitors to Rekawa and with the support of the local community endangered turtles would be better protected (Rathnayake, 2015). The developed scenario is as follows.

13.4.2.1 SCENARIO

The Rekawa beach is ~4 km long and is a prime turtle-nesting habitat in Sri Lanka. Therefore, it has been declared as a sanctuary under the Fauna and Flora Protection Ordinance (1937). The Department of Wildlife Conservation on behalf of the Government of Sri Lanka is currently working on a program to introduce conservation policies to protect turtles at Rekawa and is at the same time giving incentives to the local community. Under the plan, the Department of Wildlife Conservation will improve visitor services and other facilities to attract more visitors to the Rekawa sanctuary. In order to generate income for local communities, the Department of Wildlife Conservation will simultaneously recruit and train paid nature guides. The government will take responsibility for the safety and security of the tourists and supervise the nature guides. In terms of visitor services, the department plans to ensure there is:

1. clean toilets and available drinking water;
2. a visitor center/museum to educate tourists;
3. a cafeteria for food and snacks;
4. souvenir shops;
5. a camp site for overnight stays;
6. nature guides.

However, because of the poaching of turtles and eggs by local people, tourists often do not see the turtles and/or their nests and hatchlings. So, in order to ensure that tourists will see more turtles, turtle nests, and hatchlings, the new program will collaborate with local people to ensure that they will:

1. not collect or destroy turtle eggs;
2. not poach turtles coming to the site;
3. be engaged in the sanctuary to protect of turtles, their nests, and hatchlings;
4. be trained to protect turtles and turtle nests.

It is expected that these changes will facilitate better turtle watching opportunities for visitors and improve protection for the turtles in the sanctuary.

13.4.2.2 BID VEHICLE

The second element of the CV question is the method, or vehicle for paying for the recreational services or wildlife conservation program. A common and natural method is to link public good services with tax payments or entrance fees, but other methods such as payments on utility bills are used.

Acceptable vehicles provide a clear link, one that conveys the necessity of payment in return for the service or to establish the wildlife conservation program. In the above turtle study the payment vehicle was an entrance fee. Applied bids in the case study were USD 5, 10, 15, and 20.

13.4.2.3 QUESTIONING METHODOLOGY

The final element of a CV scenario is the method of asking questions. This part of the questionnaire presents the respondent with a given monetary amount, and encourages a response. In this study, Rathnayake (2015), the dichotomous choice approach was used and the respondents were asked, "For conserving the turtles in the Rekawa under the Scenario 1, the proposed entrance fee for entering the sanctuary is USD 5. Would you like to pay this amount?" This is a "referendum CVM" question, and the USD amounts, that is, USD 10, 15, 20, and so forth, vary across the respondents.

The dichotomous choice approach has become the presumptive method of elicitation for CV practitioners. The other three methods have been shown to suffer from incentive compatibility problems when survey respondents influence potential outcomes by revealing values other than their true WTP.

13.4.2.4 DATA ANALYSIS MODEL

People generally choose what they prefer, and where they do not, this can be explained by random factors. For example, a person may choose their preferred ice cream nine out of 10 times and on the 10th occasion they choose something else due to some random factor. The basic model for analyzing CV responses is that random utility model. Although Bishop and Heberlein (1979) employed a dichotomous question for CV responses, Hanemann (1984) constructed the basic model. Utilizing the random utility framework that McFadden (1986) developed, Hanemann (1984) rationalized responses to dichotomous CV questions, putting them in a framework that allows parameters to be estimated and interpreted.

In the CV case studies there are two choices or alternatives so that indirect utility for respondent j can be written as:

$$u_{ij} = u_i(y_j, z_j, \varepsilon_{ij})$$

where $i = 1$ is the state or condition that prevails when the CV program is implemented, that is, the final state, and $i = 0$ for status quo. The determinants of utility are y_j, the *j*th respondent's discretionary income, z_j, an m-dimensional vector of household characteristics and attributes of the choice, including questionnaire variations and ε_{ij} a component of preferences known to individual respondent but not observed by the researcher.

In the CV study, all we know is that something has been changed from the status quo to the final state (hypothetical situation). It could be a measurable attribute, for example, an indicator q could change from q^0 to q^1 so that utility for the status quo would be $u_{0i} = u(y_j, z_j, q^0 \varepsilon_{ij})$ and the utility in the final state would be:

$$u_{1i} = u(y_j, z_j, q^1 \varepsilon_{1j})$$

The probability of a yes response is the probability that the respondent thinks that he or she is better off in the proposed scenario, even with the required payment, so that $u_1 > u_0$. For respondent *j*, this probability is:

$$\Pr(yes_j) = \Pr\left(u_1\left(y_j - t_j, z_j, \varepsilon_{ij}\right)\right) > u_0(y_j z_j, \varepsilon_{0j})$$

The expected linear function for the dependent variable (y) and dependent variables (Xs) is as follows:

$$y = \beta_0 + \beta_1 X_1 + \beta_2 X_2 \ldots\ldots + \varepsilon_i$$

Sometimes we had to transform or add variables to ensure the equation was linear:

1. taking logs of y and/or the Xs;
2. adding squared terms;
3. adding interactions.

Researchers commonly use probit and logit models when the dependent variable is binary (Bishop & Heberlein, 1979; Capps & Kramer, 1985; Seller et al., 1985). If we assume that the error term u_i has a normal distribution $N(0, \sigma^2)$, in this case a probit model is applied.

In the above turtle study, to estimate the WTP, we follow the estimation approach given in Lopez-Feldman (2012). In this approach, WTP is modeled as a linear function:

$$WTP_i(z_i, u_i) = z_i\beta + u_i \tag{13.1}$$

where z_i is a vector of explanatory variables, β is a vector of parameters, and u_i is an error term. In our study, the following are the explanatory variables:

educ:	education in number of years
age:	age of respondents in years
gender:	respondents' gender (1= male and 0= female)
hhinc:	household income (LKR/USD)
marital:	marital status (1= married, 0= others)
entow:	working in tourism or environment-related field (1= yes, 0= no)
turt seen:	seen turtles (1= yes, 0= no)
nest seen:	seen turtle nesting (1= yes, 0= no)
site dum:	survey site (1= Rekawa, 0= Yala and Bundala)
rekawavi:	knowledge of Rekawa or visited Rekawa (1= yes, 0= no)
grsize:	group size

Each individual is offered a single bid value (t_i) and is expected to answer yes or no. Denote $y_i = 1$ if the answer is yes and $y_i = 0$ if the answer is no. The individual would answer yes when his or her WTP is greater than the offered bid amount ($WTP_i > t_i$). The probability of $y_i = 1$ is a function of the explanatory variables and can be written as:

$$\Pr(y_i = 1 | z_i) = \Pr(WTP_i > t_i) \tag{13.2}$$

$$\Pr(y_i = 1 | z_i) = \Pr(z_i\beta + u_i > t_i)$$

$$\Pr(y_i = 1 | z_i) = \Pr(u_i > t_i - z_i\beta) \tag{13.3}$$

In this study, the outcome is binary and we apply the probit model to analyze the data. Hence, we assume that the error term u_i has a normal distribution $N(0, \sigma^2)$. In this case, Equation (13.3) can be rewritten as:

$$\Pr(y_i = 1 | z_i) = \Phi\left(\frac{z_i\beta}{\sigma} - \frac{t_i}{\sigma}\right) \tag{13.4}$$

where $\Phi(.)$ denotes the standard cumulative normal distribution function. Note that in Equation (13.4) the probit model has t_i in addition to z_i as explanatory variables. There are two ways in which one could estimate this model. The first is to use Equation (13.4) and apply maximum likelihood estimation to solve β and σ. The other option, which we use in this study, is to estimate directly the probit model with z_i and t_i as explanatory variables, which can be estimated in STATA. In this case, we obtain estimates of β/σ and $-1/\sigma$ after estimating the probit model. For the results of the probit model, denote $\hat{\alpha} = \hat{\beta} / \hat{\sigma}$ (the vector of coefficients associated to each one of the explanatory variables) and $\hat{\delta} = 1 / \hat{\sigma}$ (the coefficient for the variable capturing the amount of bid).

The expected value of WTP can be estimated for individuals with certain characteristics or as the average of explanatory variables:

$$E(WTP | \tilde{z}) = \frac{\tilde{z}\,\hat{\alpha}}{\hat{\delta}} \tag{13.5}$$

where $\tilde{z}$ is a vector with the values of interest for the explanatory variables.

Using a statistical package (STATA) mean WTP values, margins, and WTP values under the socio-economic attributes can be estimated.

13.4.2.5 RESPONSES TO CV QUESTIONS

This section measures users' mean WTP for access to the Rekawa sanctuary given the different entrance fees (proposed bid values) if the developed scenarios are implemented. Table 13.1 shows the estimated number of visitors to Rekawa under two different scenarios and bid values. We found that by increasing the bid value as entrance fees, the number of visitors who are willing to pay for scenarios went down gradually. Furthermore, we found that compared to scenario 2, the WTP for scenario 1 drastically fell with increasing bid value. We plotted the number of expected local visitors against the proposed entrance fees, and Figure 13.1 shows the results for expected visitor arrivals under the two different scenarios and proposed bid values.

TABLE 13.1 Estimated Number of Local Visitors to Rekawa and Revenue Changes in Response to Different Scenarios and Bid Values.

Bid value (SLR)	Scenario 1		Scenario 2		Revenue change (SLR)	Revenue change (%)
	Estimated number of visitors	Estimated revenue (SLR)	Estimated number of visitors	Estimated revenue (SLR)		
0	80,392	0	95,656	0	0	0.00
60	62,075	3,724,489	84,462	5,067,748	1,343,259	36.07
75	56,987	4,274,004	79,374	5,953,077	1,679,073	39.29
100	47828	4,782,814	70,216	7,021,578	2,238,764	46.81
125	39,687	4,960,898	59,022	7,377,745	2,416,847	48.72
150	30,529	4,579,290	47,828	7,174,221	2,594,931	56.67

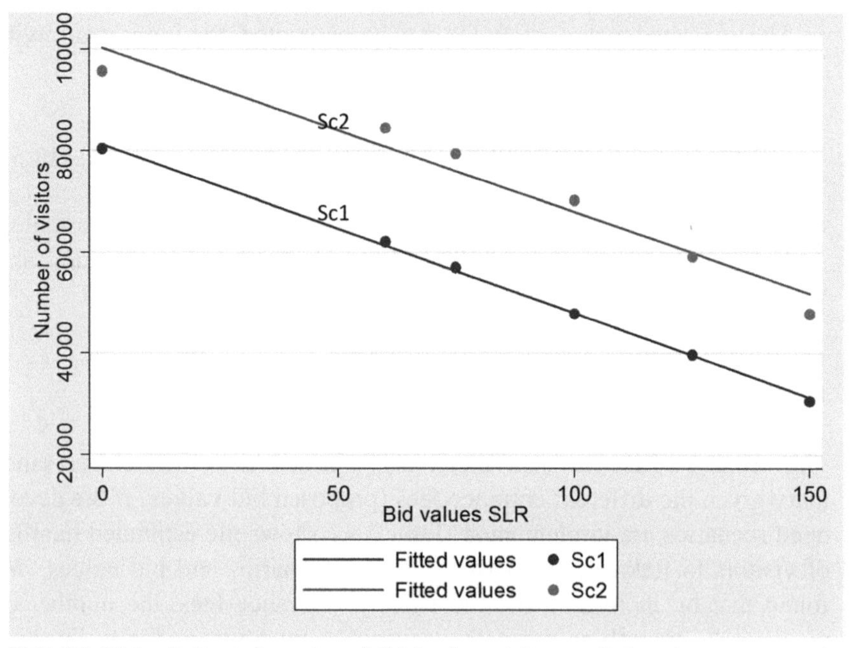

FIGURE 13.1 Estimated number of Sri Lankan visitors to Rekawain response to the different bid values of scenarios 1 and 2.

Note: *Sc1, estimated visitors to Rekawa under scenario 1; Sc2, estimated visitors to Rekawa under scenario 2.*

13.5 VISITORS' WTP

In CV studies a valuation function is estimated that relates discrete choice or WTP to variables that are expected to have an influence on the choice or on the stated WTP amount. This explorative estimation allows us to perform a test of construct and theoretical validity by determining whether choices or WTP amounts are significantly related to variables suggested by economic theory.

In order to analyze visitors' WTP for visitor facilities and/or marine turtle conservation cum visitor services, we first pooled the data from both scenarios for each site and for all visitors. Table 13.2 gives the estimated probit models that best fitted the data.

TABLE 13.2 Probit Regression Results for Each Scenario.

Variables	Scenario 1	Scenario 2
	Prob	Prob
Bidv	–0.00928***	–0.0132***
	(0.00177)	(0.00221)
Age	–0.00187	–0.0160*
	(0.00853)	(0.00912)
Gender	0.0859	0.145
	(0.186)	(0.191)
Marital	0.124	0.0214
	(0.195)	(0.195)
Educ	0.120***	0.137***
	(0.0309)	(0.0334)
Entow	0.697**	0.973***
	(0.284)	(0.340)
Hhinc	5.33e–06	3.07e–05***
	(4.35e–06)	(9.50e–06)
Grsize	–0.0181*	0.0145
	(0.0103)	(0.00956)
Site dum	–0.415**	0.233

TABLE 13.2 *(Continued)*

Variables	Scenario 1	Scenario 2
	Prob	Prob
	(0.191)	(0.219)
Turt seen	0.612***	0.401*
	(0.191)	(0.221)
Nest seen	–0.0674	0.0237
	(0.221)	(0.246)
Rekawavi	0.339	0.113
	(0.209)	(0.210)
Constant	–1.142*	–1.231*
	(0.606)	(0.686)
Observations	300	300

Notes: *Standard errors in parentheses.* $^{***}p < 0.01$, $^{**}p < 0.05$, $^{*}p < 0.1$.

As expected, the estimation parameters with regard to "bidv" (bid value) were negative in all estimations and were statistically significant at the 0.05 significance level. This means that, with a higher bid value, respondents would reduce their number of visits to Rekawa in line with the economic theory of demand, which is not supportive of the turtle conservation initiative. Conforming to *apriori* theoretical expectations, the coefficient on the "hhinc" (household monthly income) variable is positive and significant, implying that income is an important factor affecting the WTP for visitor facilities and/or turtle conservation. Respondents with higher household incomes would pay more for visitor facilities and/or the turtle conservation program. In the pooled model, the variable "educ" (years of education) was also positive and significant. Age and gender were not significant but working in the fields of tourism or the environment is positive and significant for scenario 2. This means visitors from tourism or environmental related fields are willing to pay higher entrance fees for turtle conservation. If the respondents have visited Rekawa and seen turtles, they are willing to pay, because they recognize the importance turtle conservation and the Rekawa site. Site dum (Rekawa) is not significant for WTP in all scenarios. This finding suggests that the type of recreational activity engaged in at a particular site is not an important determinant of demand for turtle conservation and improvement of visitor facilities at Rekawa. In this study, some variables are significant at the 0.05% level, while the majority of variables are significant at 0.1% for both scenarios.

The visitors' WTP is SLR 93.08 and SLR 142.61 for scenarios 1 and 2, respectively, and in terms of USD the estimated mean WTP is almost same as in Jin et al.'s (2010) study. These values were comparable with WTP values for other species from studies in Asia (Harder, 2006). We found that the mean WTP is high for scenario 2 in amounts of Rs. 40.00, because visitors are interested in turtle conservation rather than in enjoying the recreational facilities. In addition, they may feel that the initiatives mentioned in scenario 2 will help to solve the problems at Rekawa and improve turtle survival. The WTP values identified in this study were comparable to others from studies of East Asia and developed countries. Jin et al. (2010) conducted a cross country study for turtle conservation and found different mean WTP values for different cities. It can be seen that for lower income cities, such as Davao City in the Philippines and Ho Chi Minh/Hanoi in Vietnam, the mean WTP was around USD 0.30/month per household. For the relatively higher income cities, the mean WTP values of Beijing and Bangkok households were USD 1.28 and USD 1.08/month per household, respectively. In this study, compared to the above values, foreign visitors are willing to pay more for turtle conservation initiatives.

Respondents in developed countries would be willing to pay 0.24 and 0.08% of their annual per capita income to support the spotted owl (Loomis & Ekstrand, 1998) and the Minke whales in the Northeast Atlantic (Bulte & Van-Kooten, 1999), respectively. We found that local visitors were likely to pay 0.005% of their per capita income for the turtle conservation.

13.6 MARKET AND REVENUE CHANGES UNDER DIFFERENT SCENARIOS

If we implement scenario 2, visitors are willing to pay even higher entrance fees (Table 13.2). Therefore, although the expected number of visitors has decreased, revenue generation has increased gradually by increasing the entrance fee up to the mean WTP. After the mean WTP revenue will decrease. Figure 13.2 displays the increments and reductions in revenue in response to increasing bid values. This figure also shows that up to the mean WTP level revenue will increase and after that point revenue will gradually decrease. According to the calculated revenue changes, scenario 2 is highly marketable to the visitors: visitors are willing to enjoy recreational facilities as well as turtle conservation rather than just enjoying the recreational facilities described under scenario 1.

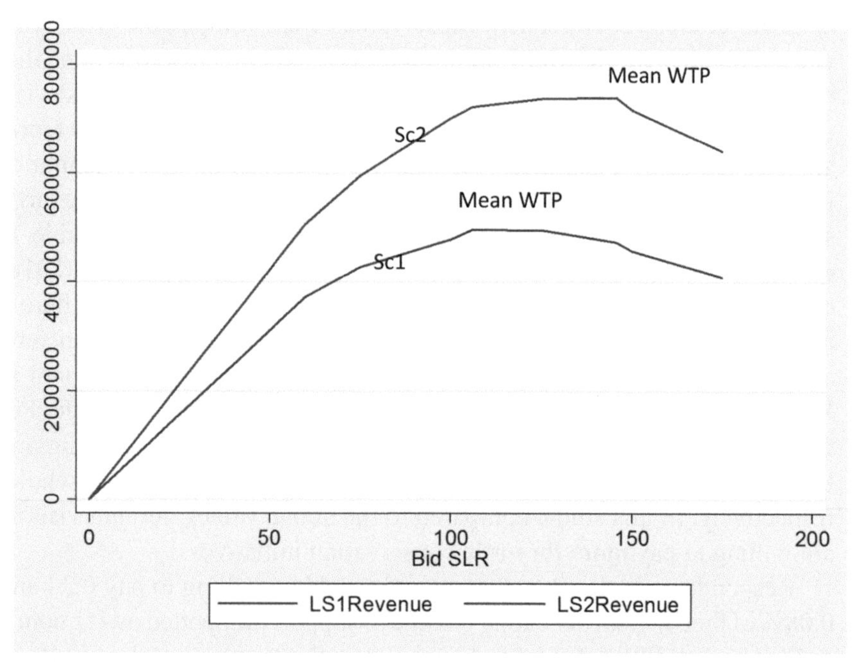

FIGURE 13.2 Revenue changes with increasing bid values in response to different scenarios for local visitors.

Note: Sc1,scenario 1 for visitors;Sc2,scenario 2 for visitors.

Table 13.3 shows the marginal effects of the proposed bid values under the different scenarios. As discussed in section 4.4.5, the variables of bid value, household income, education, visiting experience to Rekawa, and working in tourism or environment related fields are significant at the 0.05% confidence level. The marginal effects of estimated mean WTP values were 0.530 and 0.779 under scenarios 1 and 2, respectively. Therefore, we found that there is an increase in marginal effect with improving the recreational facilities and ensuring conservation.

TABLE 13.3 Marginal Effects after Probit Model Regression.

Y= Pr (prob) (mean WTP)	Scenario1		Scenario 2	
	0.530		0.779	
Variable	**Coefficient**	**Standard error**	**Coefficient**	**Standard error**
Bidv	–0.004***	0.001	–0.004***	0.001
Age	–0.001	0.003	–0.005*	0.003
Gender	0.034	0.074	0.044	0.059
Marital	0.049	0.078	0.006	0.058
Educ	0.048***	0.012	0.041***	0.010
Entow	0.257***	0.091	0.206***	0.046
Hhince			0.00001***	0.00000
Grsize	–0.007*	0.004	0.004	0.003
Site dum	–0.164**	0.075	0.069	0.065
Turt seen	0.240***	0.073	0.121*	0.067
Nest seen	–0.027	0.088	0.007	0.073
Rekawavi	0.133*	0.081	0.033	0.061

Note: Standard errors in parentheses. ***$p < 0.01$, **$p < 0.05$, *$p < 0.1$.

Under scenario 1, the calculated marginal effects for local visitors are not high. If we increase the bid value by SLR 1, the probability of acceptance of the scenario will decrease by 0.40%. The entry fee (in terms of bid value) increases by SLR 60.00, 75.00, 100.00, 125.00, and 150.00, therefore it could be found that the probability of acceptance decreases drastically with increasing entry fees. If the visitors are educated, and they have visited Rekawa, already seen turtles, and worked in tourism or environment related fields it will increase the probability of acceptance of scenario 1 by 4.8, 13.3, 24, and 25.7%, respectively.

Visitors also demonstrated that if we increase the bid value by SLR 1 under scenario 2, the probability of acceptance of the scenario will decrease by 0.40%. As with scenario 1, if the visitors are educated, the probability of acceptance of scenario 2 will increase by 4.1%. Meanwhile, if they have seen turtles and worked in tourism or environment related fields, the probability of acceptance of scenario 2 increases by 12.1 and 20.6%, respectively.

If recreational facilities are improved visitors who are educated, working in tourism or environment related fields and have visited Rekawa before are willing to pay more for both scenarios.

13.7 IDENTIFICATION OF ECONOMIC INCENTIVES FOR THE LOCAL COMMUNITY

The total number of families in the Rekawa area (GramaNiladhari Division: Rekawa West 255) is 280. The main sources of income are fishing (lagoon and fresh water fishing), carpentry, masonry, handicraft production, home garden cultivation, the coir product industry, and the tourism industry. The total population in the area is 1062, and out of that only 160 people work in either the public sectors or private sectors. The majority of people do not have access to basic necessities, such as drinking water and electricity. The poverty percentage in the area is about 33%, and their daily household income is less than Rs.100.00. There are six hotels sited close to Rekawa beach; some are "star hotels," in which local people work. In the fishing industry, a few families benefit from lagoon fishing. The local community is unable to compete with village level rich people in fishing, because the rich people operate the fishing activities within the area. In addition a few (about 17 families) directly benefit from working as so called "nest protectors," preventing others from being employed in turtle conservation. These nest protectors also steal turtle eggs to meet their daily needs and the local community in general is involved in turtle egg collection and occasionally in poaching turtles. During the field studies we found that some local youth earn up to Rs.1000.00 working as nest protectors. The egg collectors earn up to Rs.300.00 stealing turtle eggs. The majority of the local community lives under the poverty line and they are unemployed or underemployed. They do not have the opportunity to work as nest protectors or guides in the "existing turtle watching" economy, because the existing nest protectors maintain a monopoly on the positions as guides or nest protectors, while at the same time stealing turtle eggs. Though some local and international non-government organizations provide funds, they have not established the necessary infrastructure and visitor facilities for "turtle watching."

Rekawa is situated within the tourism destination of Tangalle beach, which could be an advantage when promoting turtle watching. The lack of opportunity for the local community to get involved in "turtle watching" is the major threat to the conservation effort. The consumption of turtle eggs, lack of knowledge about turtles and turtle nesting, and negative attitudes toward the protection of turtles could be considered as other significant threats to the turtles. The government should be more involved in policymaking because the beach is a government property as declared under Fauna and Flora Protection Ordinance. According to the results and information obtained we propose the following the policy decisions to provide economic incentives to the local community to conserving turtles.

1. To streamline and promote "turtle watching" at Rekawa.
2. To propose a proper incentive/compensation scheme for the local community to get them involved in turtle conservation.

At present, annual earnings from foreign visitors is SLR 3.20 million. The estimated mean WTP values for Sri Lankan visitors were SLR 106.05 and 145.47 for scenarios 1 and 2, respectively. Under scenarios 1 and 2, the turtle watching initiative could earn SLR 4.96 million and SLR 7.40 million, respectively, by selling entry tickets to Sri Lankans. Therefore, this would represent a 154.91 and 230.98% annual revenue increase under scenarios 1 and 2, respectively. Then the sanctuary authority, DWC, would be able to provide more economic incentives to the local community.

One of the major concerns expressed by the local community is that they will be absorbed for the turtle conservation initiatives. In fact, this is a major concern for administrators, policymakers, and local community leaders. Given high education levels and the positive attitudes of the local community toward tourism, coupled with recognition of the importance of turtle conservation, there is every possibility of providing temporary employment for the local community in infrastructure construction. Following that they could have permanent employment operating visitor facilities and working as qualified interpreters/nature guides and nest protectors.

In this study we found that some local businesses are directly related to tourism, for example, batik textile industry, coir item production, local sweet production, coir based ornaments production, and other handicrafts. Policymakers should promote these local businesses by providing the necessary education and training in order to provide economic incentives to the local community while at the same time conserving the turtles.

Furthermore, local banks should provide the local community with loan schemes to improve their businesses.

Local hoteliers, restaurant operators, and boutique owners are prepared to provide visitor facilities, including accommodation and meal services, if there is government involvement in the tourism operation at Rekawa. They believe government involvement is required in tourism initiatives, because government involvement in tourism is more widely accepted by the general public compared to direct local community involvement. They believe such involvement would draw more visitors to the beach. Other government officials such as divisional secretaries, policemen, forest officers, and social service officers also mentioned that DWC should be involved in policymaking, law enforcement, and managing the tourism initiatives, because Rekawa is a declared sanctuary by DWC under the Fauna and Flora Protection Ordinance and powers are provided for in that ordinance for the conservation of turtles. However, under their legal framework the local community should get involved in working as nest protectors, nature guides, and visitor center staff. Policymakers should widen opportunities so the whole local community can get involved in tourism initiatives. In the preliminary and individual discussions, the existing nest protectors or nature guides also expressed their acceptance of the proposals mentioned in scenario 2.

The resulting mean WTP will help to determine the entry fee for fee-levying turtle watching at the Rekawa sanctuary. In addition, we could identify the most suitable management scenario in this study to provide the required funds to establish visitor facilities. If the government encourages private–public partnership with local investors, the government will not be able to allocate more funds to establish these facilities. If we pay a portion of the entry fee to the local community, we will be able to create employment through tourism initiatives at the Rekawa sanctuary. The amount of salary/allowance that is given to young employees is determined on their minimum daily living expenditure.

1. Educated youths are recruited as nature guides/interpreters.
2. Young energetic youth are recruited as nest protectors.

Policymakers could generate the following employment opportunities through the improvement of visitor facilities/services at the Rekawa sanctuary. We identified the potential for these opportunities in our initial discussions with the local community, community leaders, and hotel and boutique owners.

3. Skilled youth can produce handicrafts.
4. Skilled youth/local community can operate restaurants.
5. Skilled local community can maintain clean toilet facilities.
6. Local women can sell their handicrafts/souvenirs to the tourists.
7. Local community can operate organized home stay program (local community provides accommodation for guests) and camping facility.

13.8 CONCLUSION AND POLICY RECOMMENDATIONS

Policymakers could secure the local community's involvement in turtle conservation through the "turtle watching" tourism initiative. According to the results, the WTP values recorded for scenario 2 are considerably higher than that for scenario 1. This indicates that scenario 2 could be implemented at Rekawa sanctuary as a policy decision. If we implement scenario 2, which will provide both opportunities for recreation and turtle conservation, there will be a 230.98% annual revenue increase (SLR 7.40 million).

Furthermore, based on the results and conclusions in this study, we found that there is a potential to promote a "turtle watching" tourism initiative which would help conserve turtles at Rekawa sanctuary. The following policy directions would support both turtle conservation and the provision of economic incentives to the local community.

1. Streamlining and promotion of turtle watching at Rekawa.
2. Introduction of new a fee structure for turtle watching.
3. Proper incentive/compensation scheme for the local community to encourage them to get involved in turtle conservation.
4. Establishment of a village level welfare fund to improve the basic infrastructure in Rekawa village.

Therefore, this case study has showed that economic valuation could be useful in policy formulation in relation to wildlife tourism.

13.9 ISSUES RELATED TO CVM

Although CVM has become one of the most popular methods used by environmental and resource economists to value environmental goods, Hausman (1993), and Cummings and Harrison (1994) criticized the fact that the technique remains controversial. Boyle et al. (1993),and Foster and Mourato (2003) showed that there is a significant body of evidence to suggest that CV estimates do not exhibit sensitivity to scope. With respect to a specific endangered species, how should funds be collected to support their conservation? Is mandatory payment more effective than voluntary contribution?

In addition, in the main survey also we found several limitations and biases in the CV, those described by Shavell (1993). That individuals may not be able to estimate or even to understand the values or harms about which they were asked was the major limitation in our study. To overcome this issue, we encouraged discussion and to understood the respondents' scientific, economic knowledge and attitudes to turtle conservation. Following the discussion we tried to obtain correct answers for WTP for the given scenario and bid value.

Another limitation was that individuals may misrepresent their benefits believing, correctly or not, that the CV results would influence public or private decisions in some manner. Therefore in this study, we had to build up trust with the respondents to obtain more accurate answers.

That individuals may lack incentives to answer carefully is another limitation. In the front page of our questionnaire, and also during our questionnaire surveys, we described the turtles and their importance without emphasizing that the turtles are the most important animals. Therefore, our interviews encouraged them to answer carefully.

CV responses may depend significantly on how questions are posed, which is another criticism of CV. We pre-tested our questionnaire several times and identified how questions are posed. In our questionnaire, we located the questions on WTP for the proposed bid value and scenario at the end of the questionnaire. This helped us to collect more accurate data.

Starting point bias, or "anchoring," is another bias in CVM, but it was not found in this study because we gave different bid values to respondents. In the "bidding game" an initial valuation figure, presented to the respondent, may be taken as an indication of the "normal" level of value or payment, and that later valuation figures may be drawn in the direction of this amount. This problem is greater when the respondent is less familiar with the object to be valued and with the valuation procedure.

We found informational biases in interviewing the respondents in this study, so that respondents provided information differently with different interviewers. Therefore, we employed one female and one male interviewer to interview the respondent to get more accurate answers for the scenarios and bid values.

Mental account or scope bias was also present. Individuals have a particular "account" to allocate to goods and services, and they do not like to pay more than this amount. Embedding is another bias in CVM. More comprehensive goods are valued nearly the same as those that are less comprehensive. This may have a theoretical basis in strong substitutability, but may else be related to the mental account issue.

In this study the bid vehicle was the entrance fee and all visitors were familiar with proposed entrance fees as bid values so that we were able to minimize vehicle biases. Although, individuals have preferences for particular vehicles.

13.10 NOAA GUIDELINES FOR CVM STUDIES

The "Exxon Valdez" incident, a large oil spill off the Alaska coast in 1989, was the background for the National Oceanic and Atmospheric Administration (NOAA) panel to review the CVM. In that case, WTP data obtained from CVM studies were brought to court. These studies were contested, and the entire CVM was seriously questioned.

In 1993, NOAA introduced the following guidelines, which are met by the best CV surveys and need to be present in order to assure reliability and usefulness of the information that is obtained. The guidelines are designed to shape CV questions in light of the shortcomings of the CV study of the Exxon Valdez oil spill. All parts are obviously not relevant for all CV settings, but the general sense of careful and thorough survey design and testing is relevant. These guidelines help to minimize or solve the issues related CV studies.

13.10.1 CONSERVATIVE DESIGN

Generally, when aspects of the survey design and the analysis of the response are ambiguous, the option that tends to underestimate WTP is preferred. A conservative design increases the reliability of the estimate by eliminating extreme responses that can enlarge estimated values widely and implausibly.

1. **Elicitation format:** The WTP format should be used instead of the compensation required because the former is the conservative choice.
2. **Referendum format:** The valuation question should be posed as a vote on referendum.
3. **Accurate description of the program or policy:** Adequate information must be provided to respondents about the environmental program that is offered. It must be defined in a way that is relevant to damage assessment.
4. **Pretesting of photographs:** The effects of photographs on subjects must be carefully explored.
5. **Reminder of substitute commodities:** Respondents must be reminded of substitute commodities, such as other comparable natural resources or the future state of the same natural resource. This reminder should be introduced forcefully, and directly prior to the main valuation question to assure that respondents have the alternatives clearly in mind.
6. **Adequate time lapse from the accident:** The survey must be conducted at a time sufficiently distant from the date of the environmental incident that respondents regard the scenario of complete restoration as plausible. Questions should be included to determine the state of subject beliefs regarding restoration probabilities. This guideline is especially relevant for natural resource accidents but may not be relevant for many other more mundane types of studies.
7. **Temporal averaging:** Time dependent measurement noise should be reduced by averaging across independently drawn samples taken at different points in time. A clear and substantial time trend in responses would cast doubt on the "reliability" of the findings. This guideline pertains to natural resource accidents that have a high public awareness such as oil spills.
8. **"No-answer" option:** A "no-answer" option should be explicitly allowed in addition to the "yes" and "no" vote options on the main valuation (referendum) question. Respondents who choose the "no-answer" option should be asked to explain their choice.

Answers should be carefully coded to show the types of responses. For example: (a) rough indifference between a yes and a no vote; (b) inability to make a decision without more time and more information; (c) preference for some other mechanism for making this decision; and (c) bored by this survey and anxious to end it as quickly as possible. Subsequent research has concluded that no-answer responses are best grouped as "no."

9. **Yes/no follow-ups:** Yes and no responses should be followed up by the open ended question: "Why did you vote yes/no?" Answers should be carefully coded to show the types of responses, for example: (a) It is (or is not); (b) Do not know; or (c) The polluters should pay.
10. **Cross-tabulations:** The survey should include a variety of other questions that help to interpret the responses to the primary valuation question. The final report should include summaries of WTP broken down by these categories. Among the items that would be helpful in interpreting the responses are: income, prior knowledge of the site, prior interest in the site (visitation rates), attitudes toward the environment, attitudes toward big business, distance to the site, understanding of the task, belief in the scenarios, ability/willingness to perform the task.
11. **Checks on understanding and acceptance:** The above guidelines must be satisfied without making the instrument so complex that it poses tasks that are beyond the ability or interest level of the majority of participants.

This chapter discussed the CVM method and its application as an economic valuation method in the wildlife tourism sector. Economic values were useful for shaping policy decisions in tourism management, marketing and non-marketed tourism products and for conserving natural resources. There were limitations/issues found in the CVM, and the NOAA guidelines would be helpful for minimizing or solving the issues affecting CVM studies.

13.11 ACKNOWLEDGMENT

The turtle study was funded by South Asian Network for Development and Environmental Economics (SANDEE).

KEYWORDS

- **willingness to pay**
- **willingness to accept**
- **hypothetical market**
- **bid value**
- **wildlife**

REFERENCES

Bishop, R. C.; Heberlein, T. A. Measuring Values of Extra Market Goods: Are Indirect Measures Biased? *Am. J. Agric. Econ.* **1979,** *61,* 926–930.

Bulte, E. H.; Van-Kooten, G. C. Marginal Valuation of Charismatic Species: Implications for Conservation. *Environ. Resour. Econ.* **1999,** *14,* 119–130.

Bowker, J. M.; Cordell, H. K.; Johnson, C. Y. User Fees for Recreation Services on Public Lands: A National Assessment. *JPRA.* **1999,** *17* (3), 1–14.

Boyle, K. J.; Welsh, M, P.; Bishop, R. C. The Role of Question Order and Respondent Experience in Contingent Valuation Studies. *J. Environ. Econ. Manage.* **1993,** *25,* 80–89.

Capps, O.; Kramer, R. A. Analysis of Food Stamp Participation Using Qualitative Choice Models. *Am. J. Agric. Econ.* **1985,** *67* (1), 49–50.

Clawson, M.; Knetsch, J. L. *Economics of Outdoor Recreation;* Johns Hopkins Press: London, 1966.

Carlsson, F.; Johansson-Stenman, O. Willingness to Pay for Improved Air Quality in Sweden. *Appl. Econ.* **2000,** *32* (6), 661–669.

Clinch, J. P.; Murphy, A. Modeling Winners and Losers in Contingent Valuation of Public Goods: Appropriate Welfare Measures and Econometric Analysis. *Econ. J.* **2001,** *111* (470), 420–443.

Cummings, R.; Harrison, G. Was the Ohio Court Well Informed in its Assessment of the Accuracy of the Contingent Valuation Method? *Nat. Resour. J.* **1994,** *34,* 1–36.

Davis, D.; Tisdell, C. Tourist Levies and Willingness to Pay for a Whale Shark Experience. *Tour. Econ.* **1998,** *5* (2), 161–174.

Eagles, P. F. J.; McCool, S. F.; Haynes, C. D. *Sustainable Tourism in Protected Areas: Guidelines for Planning and Management;* IUCN: Cambridge, UK, 2002.

Foster, V.; Mourato, S. Elicitation Format and Sensitivity to Scope. *Environ. Resour. Econ.* **2003,** *24,* 141–160.

Hanemann, W. M.; Loomis, J.; Kanninen, B. Statistical Efficiency of Double-Bounded Dichotomous Choice Contingent Valuation. *Am. J. Agric. Econ.* **1991,** *73* (4), 1255–1263.

Hanemann, W. M. Welfare Evaluations in Contingent Valuation Experiments with Discrete Responses. *Am. J. Agric. Econ.* **1984,** *66* (3), 332–341.

Harder, D. *Willingness to Pay for Conservation of Endangered Species in the Philippines: The Philippines Eagle*; EEPSEA Research Report: Singapore, 2006.

Hausman, J. A. *Contingent Valuation: A Critical Assessment;* North-Holland: Amsterdam, Netherlands, 1993.

Jin, J. J.; Indab, A.; Nabangchang, O.; Thuy, T. D.; Harder, D.; Subade, R. F. Valuing Marine Turtle Conservation: A Cross-Country Study in Asian Cities. *Ecol. Econ.* **2010,** *69,* 2020–2026.

Kerr, G. N.; Manfredo, M. J. An Attitudinal Based Model of Pricing for Recreational Services. *J. Leis. Res.* **1991,** *23* (1), 37–50.

Laarman, J. G.; Gregersen, H. M. Pricing Policy in Nature-Based Tourism. *Tour. Manag.* **1996,** *17*(4), 247–254.

Lee, C. Valuation of Nature-Based Tourism Resources Using Dichotomous Choice Contingent Valuation Method. *Tour. Manage.* **1997,** *18* (8), 587–591.

Leuschner, W. A.; Cook, P. S.; Roggenbuck, J. W.; Oderwald, R. G. A Comparative Analysis for Wilderness User Fee Policy. *J. Leis. Res.* **1987,** *19* (2), 101–114.

Loomis, J. B.; Ekstrand, E. Alternative Approaches for Incorporating Respondent Uncertainty when Estimating Willingness to Pay: The Case of Mexican Spotted Owl. *Ecol. Econ.* **1998,** *27,* 29–41.

Lopez-Feldman, A. *Introduction to Contingent Valuation Using STATA*; MPRA Paper No. 41018, Posted 4 September 2012, 19(36). Retrieved from http://mpra.ub.uni.muenchen.de/41018/.

McFadden, D. The Choice Theory Approach to Market Research. *Market. Sci.* **1986,** *5* (4), 275–297.

Mitchell, R. C.; Carson, R. T. *Using Surveys to Value Public Goods: The Contingent Valuation Method;* Resources for the Future: Washington DC, 1989.

More, T.; Stevens, T. Do User Fees Exclude Low-Income People from Resource-Based Recreation? *J. Leis. Res.* **2000,** *32* (3), 341–357.

Rathnayake, R. M. W. *Estimating Demand for Turtle Conservation at Rekawa Sanctuary in Sri Lanka;* SANDEE: Working Paper No. 92, Nepal, 2015.

Reiling, S. D.; Cheng, H.; Trott, C. Measuring the Discriminatory Impact Associated with Higher Recreational Fees. *Leis. Sci.* **1992,** *14* (2), 121–137.

Reiling, S. D.; Criner, G. K.; Oltmanns, S. E. The Influence of Information on Users' Attitudes toward Campground User Fees. *J. Leis. Res.* **1988,** *20* (3), 208–217.

Schroeder, H. W.; Louviere, J. Stated Choice Models for Protecting the Impact of User Fees at Public Recreation Sites. *J. Leis. Res.* **1999,** *31* (3), 300–324.

Seller, C.; Stoll, J. R.; Chavas, J. P. Validation of Empirical Measures of Welfare Change: A Comparison of Nonmarket Techniques. *Land Econ.* **1985,** *61* (2), 156–175.

Shavell, S. *Contingent Valuation of the Nonuse Value of Natural Resources: Implications for Public Policy and the Liability System;* Emerald Group Publishing Limited: UK, 1993; Vol. 220.

Vogt, C. A.; Williams, D. R. Support for Wilderness Recreation Fees: The Influence of Fee Purpose and Day Versus Overnight Use. *JPRA.* **1999,** *17* (3), 85–99.

Whitehead, J.; Haab, T.; Huang, J. C. Measuring Recreation Benefits of Quality Improvements with Revealed and Stated Behavior Data. *Resour. Energy Econ.* **2000,** *22,* 339–354.
Williams, D. R.; Vogt, C. A.; Vitterse, J. Structural Equation Modeling of Users' Response to Wilderness Recreation Fees. *J. Leis. Res. Resour. Energy Econ.* **1999,** *31* (3), 245–268.
World Conservation Union (IUCN) 2005; Marine Turtle Conservation Strategy and Action Plan for Sri Lanka, IUCN: Sri Lanka, 2005.

CHAPTER 14

RELATIONSHIP BETWEEN TOURISTS' EXPECTATION AND PERCEPTION OF WILDLIFE TOURISM AREAS: EVIDANCE FROM WEST BENGAL, INDIA

DEBASISH BATABYAL[1] and NILANJAN RAY[2*]

[1]*Pailan School of International Studies, Kolkata, West Bengal, India*

[2]*Netaji Mahavidyalaya, Kalipur, West Bengal, India*

[*]*Corresponding author. E-mail: nilanjan.nray@gmail.com*

CONTENTS

ABSTRACT

The separate bio-geographic region the Sundarbans is situated in India and Bangladesh. This large estuarine forest has immense potentiality to attract allocentric and midcentric tourists from around the world. But all the components of tourism development are not found to be functioning well in a cohesive manner. Apart from the unique natural settings, the standard and range of services in the form of accommodation, transportation, and public amenities including security need an extensive assessment. This assessment will certainly contribute to the destination development and its marketing. This chapter is an effort to analyze and interpret the tourist gap through expectation–perception measurement. This will also introduce a new dialog, debate, and direction of interpreting tourism-marketing strategies with its sustainable development orientation.

> **"…. there are many unexplored places for wild-life lovers, ornithologists, adventurers in Indian Sundarban"**

14.1 INTRODUCTION

The Sundarbans is the largest single block of tidal halophytic mangrove forest in the world. It has an area of ~10,000 km^2 of which 60% is located in Bangladesh and the remainder is in India. It is a UNESCO world heritage site as well. The Sundarban National Park is a national park, tiger reserve, and a biosphere reserve located in the Sundarbans delta in the Indian state of West Bengal. Sundarbans south, east, and west are three protected forests in Bangladesh. This region is densely covered by mangrove forests and is one of the largest reserves for the Bengal tiger. The history of the area can be traced back to 200–300 AD. Ruins of a city built by Chand Sadagar have been found in the Baghmara Forest Block. During the Mughal period, the Mughal kings leased the forests of the Sundarbans to nearby residents. The Sundarbans forest lies in the vast delta on the Bay of Bengal formed by the super confluence of the Padma, Brahmaputra, and Meghna rivers across southern Bangladesh. The seasonally flooded the Sundarbans freshwater swamp forests lie inland from the mangrove forests on the coastal fringe. The forest covers 10,000 km^2 of which about 6000 km^2 are in Bangladesh. It became inscribed as a UNESCO world heritage site in 1997. The area of the Indian part of Sundarbans is estimated to be about 4110 km^2, of which about 1700 km^2 is occupied by waterbodies in the forms of river, canals, and creeks of width varying from a few meters to several kilometers. The Sundarbans is intersected by a complex network of tidal waterways, mudflats, and small islands of salt-tolerant mangrove forests. The interconnected network of waterways makes almost every corner of the forest accessible by boat. The area is known for the eponymous Royal Bengal Tiger (Pantheratigristigris), as well as for numerous fauna including species of birds, spotted deer, crocodiles, and snakes. The fertile soils of the delta have been subject to intensive human use for centuries, and the eco-region has been mostly converted to intensive agriculture, with few enclaves of forest remaining. The remaining forests, taken together with the Sundarbans mangroves, are important habitat for the endangered tiger. Additionally, the Sundarbans serves a crucial function as a protective barrier for the millions of inhabitants in and around Khulna and Mongla against the floods that result from the cyclones. The Sundarbans has also been enlisted among the finalists in the new seven wonders of nature. A total 245 genera and 334 plant species were found in the delta. Differences in vegetation have been explained in terms of freshwater and low salinity influences in the Northeast and variations in drainage and siltation. The Sundarbans has been classified as a moist tropical forest, comprising primary colonization on new accretions to more mature beach forests. Historically vegetation types have been recognized in broad correlation with varying degrees of water salinity, freshwater flushing, and physiography.

The remote island communities that surround the park depend it for fish, honey, timber, and fuel wood, and have few alternate livelihood options. In recent years, eco-tourism has gained momentum. In 2006–2007, the park saw 75,000 visitors, an 18% increase over the previous year. Thus, it is important to understand whether tourism is augmenting livelihood of local people and reducing their forest dependence. The Sundarbans—the place that derives its name from Sundari trees has a wide variety of trees that typically thrive in estuarine conditions of high salinity, minimal soil erosion, and frequent inundation by high tides. The tidal rivers and mangrove forests provide habitats suitable for animals inhabiting tidal swamp areas. Numerous aquatic and semi-aquatic animals inhabit these forests, with their life systems being interlinked with the animals thriving in the land areas. The Sundarbans is home to an amazing variety of wild animals including spotted deer, monkeys, wild pigs, herons, white-bellied eagles, kingfishers, and about 270 Royal Bengal tigers. There have been occasional incidents of tigers of the Sundarbans attacking and killing human beings. It is believed that the uniqueness of the habitat and the lack of suitable prey have resulted in such unique behavioral trends of the Sundarbans tiger. Apart from hosting tigers, the reserve is also home to diverse aquatic and reptile life forms including the endangered Olive Ridley sea turtle, Green turtle, Hawk's Bill turtle, hard-shelled Batgur Terrapin, king cobra, pythons, chequered killback, estuarine crocodile, monitor, and lizards like the Salvator lizard to name a few. A number of trans-Himalayan migratory birds can also be spotted at the Sundarbans. The main areas for wildlife tourism are Sajnekhali Bird Sanctuary, Sudhanyakhali Watch Tower, Dublar Char Island, and Hiron Point. Sajnekhali is the home to egrets, herons, and several other species of birds. In Sajnekhali Bird Sanctuary, tourists can spot several colorful species of kingfisher, plovers, sandpipers, whimbrels, white-bellied sea eagle, lap-wings, and curfews. Apart from being a breeding colony of herons, the bird sanctuary also houses a visitor's center where tourists can see a crocodile enclosure and a shark pond along with a turtle hatchery. In addition, this sanctuary also has a Mangrove Interpretation Centre.

The Bhagabatpur Crocodile Project has, of late, emerged as an important tourist destination of the Sundarbans. This is the only crocodile project in West Bengal and is located adjacent to the Lothian Island and on the bank of the Saptamukhi Estuary. The dense mangrove forest at the confluence of Saptamukhi river system has immense natural beauty to attract tourist all throughout the year. This hatchery of estuarine crocodile and Batagur Baska species of tortoise has crocodiles of varying ages.

From Sudhanyakhali Watch Tower, most of the tigers in the reserve can be sighted. Some other wildlife species like axis deer, wild boars, and crocodiles may also be seen from this watch tower. This watch tower has a capacity to host 25 persons at a time. There is a sweet water pond where animals come to drink water. Behind the pond are stretches of land bereft of any vegetation where one can sight animals from a distance.

Hiron Point is another tourist spot in the Sundarbans and is also called the world heritage state. Tourists here can enjoy the beauty of wild nature, especially walking and running of dotted deer. There are also two other heritage sites in the Sundarbans: one is Kochikhali and the other is Mandarbaria where one can find deer and birds. If one is lucky enough then one can see the great Royal Bengal tiger too; but for sure one can at least see the stepping of great Royal Bengal tiger here and there in these spots. Besides, there are many unexplored places for wildlife lovers, ornithologists, and adventurers in Indian Sundarbans.

14.2 RESEARCH CONTEXT AND OBJECTIVES

Conventionally, tourism in West Bengal is confined to leisure, recreation, and business. Though the state does not lack national parks, biosphere reserve sanctuaries, heritage forests, and other reserve areas, this part has not come to light with necessary details and variety. Again, even the people of the state and its neighbors lack adequate information and interest. As the market for wildlife and adventure has been increasing, the potentiality for the same in the largest estuarine forests in the world needs to be assessed. As one of the revenue-generating sources for the local community and industry, tourism has been revisited along with certain strategic issues with respect to destination management. The objectives of the studies are

1. to study the growth, development and emergence of wildlife tourism in the Sundarbans, India, and
2. to measure the expectation and perception gaps of tourists to list down measures for future development and to supplement financial and marketing plans for supply and demand.

To measure or quantify the objectives above, following is the main hypothesis of the study.

H_0 There is no difference between tourists' expectations and perception.

14.3 LITERATURE REVIEW

Destination market segments varying from visitor demographics and travel characteristics to important supply led aspects of a destination. Kamra (2001) has opined that what less developed countries and regions need and require most is a fundamental economic development that addresses the poverty and pollution suffered by many in the less developed world. However, reflecting the evolution of development theory from economic-growth-based modernization models to the alternative/sustainable paradigm, the tourism development debate has similarly moved from support of the positive contribution of tourism to economic development to criticism of tourism development. Nicolau and Más (2006) explain how the existence of strong heterogeneous tourism demand is introducing a wide and diverse range of market segmentation for the choice of a destination with an increasing emphasis on relationship marketing, as the analysis of tourist destination choice represents one of the most fruitful lines of investigation in tourism studies and distinguishes various approaches to the definition of tourist destination. Alternatively, this study presents the innovation of identifying decision-making processes in an individual-by-individual, tourist-by-tourist manner. To achieve this, the authors propose a segmentation of the tourism market based on revealed preferences toward a destination. These revealed preferences have the two-fold implications that allowed to form groups of tourists with similar preferences or to treat them individually. The second section reviews the analysis of choice in tourism, in which the authors state the importance of studying the choice behavior of tourists, through revealed preferences and compared them with stated preferences with a viable literature survey of destination choice and related attributes. The third section presents the research design, including the details of the methodology applied and the sample and data used. The fourth section shows the results obtained, both from the estimation of the utility function for each tourist and from the segmentation analysis. In the fifth section, the implications of management and its future lines of research are discussed.

In her study, "India's Image as a Tourist Destination - a Perspective of Foreign Tourists" Chaudhary (2000) introduced a set of 152 questionnaires with a purpose of determining pre- and post-trip perceptions of foreign tourists about India as a tourist destination. A gap analysis between expectations and satisfaction levels was used to identify strengths and weaknesses of India's tourism-related image dimensions so that necessary efforts can be made to ensure that tourists' expectations are met. It was observed that India is rated highly for its rich art forms and cultural heritage. However, irritants like cheating, begging, unhygienic conditions, and lack of safety dampen the spirits of tourists. India can be positioned on the world map only after these hygiene factors are improved along with other motivators. In their study "The Destination Attribute Management Model: an Empirical Application to Binton, Indonesia," Litvin and Ling (2001) have explained how marketing a destination requires an understanding of vacationer perception. The gap analysis provided by normal pre- and post-visit surveys represents a good starting point in evaluating how the visitor feels about the destination, but yields only part of the story. This chapter suggests that two additional gaps should be evaluated: the gap between the general public and the purchaser, and the gap exists between one-time and repeat visitors. Using Bintan Resorts, Indonesia, a relatively new self-contained resort near Singapore, this chapter provides a tool for evaluating these gaps, and at the same time provides an interesting view of Bintan, a destination that has set its sights upon becoming the "Hawaii of the Orient." In his book "Tourist Behavior: Themes and Conceptual Schemes," Pearce (2000) has addressed tourist behavioral pattern, their social roles and individual characteristics, motivational pattern, choice for a destination, social contact, on-site experiences, and tourist's reflections on experiences. He has opined how tourists and industry leaders care for tourists' experience. The above topics are found to be broadly outlining a travel career pattern approach to motivation, multi-attribute view of destination image, socially embedded view of destination choice, discursive view of constructing relationship, benchmarking view of satisfaction and a social representation and mindful approach to visitor experiences and learning (Moscardo, 2008).

In the chapter, "The causes of deterioration of Sundarban mangrove forest ecosystem of Bangladesh: conservation and sustainable management issues," Rahman (2010) exhibited the causes and effects of damage of natural and anthropogenic resources. The growing human population with few alternative livelihood opportunities has been dealt with as serious threats to the Sundarbans mangrove forest. They also added that the rapidly expanding shrimp farming industry is a significant threat to the mangrove forests of Bangladesh. Due to illegal cutting, encroachment of forest areas, and illegal poaching of wildlife, the mangrove forest is found to be losing biodiversity in an alarming rate. This forest ecosystem also has become vulnerable to pollution, which may have changed the ecosystem's biogeochemistry. Further threats arise from global climate change, especially sea level rise. This study also recommends the application of sustainable management strategies covering needs for an advanced silvicultural system, improvement of scientific research as well as conservation measures.

14.4 DATA COLLECTION AND METHODOLOGY

To fulfill the above objectives, this study is based on both primary data and secondary data. For fulfillment of the objectives this study, observation method and survey method are used as the technical tools. The instrument of survey included questions on different influencing factors of tourism. Information about the profile of tourism industry including the tourism units (i.e., hotel, guesthouse, etc.), room numbers, and the number of local persons engaged as well as the profile of tourists were the main elements of the survey. The sample size was restricted to 120 respondents. Researchers distributed questionnaires to 120 respondents and 100 questionnaires were found suitable for deriving results. Respondents were asked what about their expectations on the first day of visit at in the Sundarbans and after three days researchers again distributed those particular respondents to fill same questionnaires to find whether their perceived services are up to the mark to their expectation or not. The questionnaire was prepared in both Bengali and English with forward translation keeping in mind the same meaning and numbers. Any one of the two versions was given to the respondent asking their choice. Collected data were analyzed by Will-coxon Paired Ranked Test in SPSS 17.

To test the above hypothesis statement whether null hypothesis is accepted or rejected, researcher has applied Wilcoxon Signed-Rank Test for Paired Samples through SPSS package.

The test statistics is given by

$$Z = \frac{T - n(n-1)/4}{\sqrt{n}\ (n+1)(2n+1)/24}$$

For a given level of significance $\acute{\alpha}$, the absolute sample Z should be greater than the absolute $Z_{\acute{\alpha}/2}$ to reject the null hypothesis. For a one-sided upper tail test the null hypothesis is rejected if the sample Z is greater than $Z_{\acute{\alpha}}$ and for a one-sided lower tail test the null hypothesis is rejected if the sample Z is less than $-Z_{\acute{\alpha}}$

To test the hypothesis in the study, the statement whether null hypoth-esis is accepted or rejected, researcher has applied Wilcoxon Signed-Rank Test for Paired Samples through SPSS package.

Table 14.1 depicts about the descriptive statistics of VAR00001 and VAR00002 of 100 numbers of respondents. From this table VAR00001 denotes before visit and VAR00002 denotes after visit of tourists at the Sundarbans.

TABLE 14.1 Descriptive Statistics.

	N	Mean	Std. dev.	Minimum	Maximum
VAR00001	100	86.5500	4.4571	75.00	98.00
VAR00002	100	85.1200	4.9793	70.00	98.00

Table 14.2 denotes test of Wilcoxon Signed Ranks, which depicts that three patterns of ranks (a) negative ranks which denote VAR00002 < VAR00001, (b) positive ranks which denote VAR00002 > VAR00001, and (c) ties which denote VAR00001 = VAR00002.

TABLE 14.2 Mean Rank and Sum of Rank of Wilcoxon Signed Ranks Test.

		N	Mean rank	Sum of ranks
VAR00002 - VAR00001	Negative ranks	51[a]	61.37	3130.00
	Positive ranks	49[b]	39.18	1920.00
	Ties	0[c]		
	Total	100		

[a]Negative ranks which denote VAR00002 < VAR00001.
[b]Positive ranks which denote VAR00002 > VAR00001.
[c]Ties which denote VAR00001 = VAR00002.

In Table 14.3, the *Z*-value is –2.085 and has a *p*-value of .037 which is greater than the calculated value 1.96 at 5% level of significance there is not enough evidence to accept the null hypothesis, thereby indicating that there are different aspect gaps which are not standard up to the mark of tourist's demand.

TABLE 14.3 Test Statistics of the Wilcoxon Signed Ranks Test.

	VAR00002 – VAR00001
Z	–2.085[a]
Asymp. Sig. (two-tailed)	.037

[a]*Based on positive ranks.*

Based on an extensive literature survey of similar destinations following 16 issues are chosen and addressed for the results orienting to future effects. The results presented in Table 14.4 depicted that the infrastructure variables have a significant contribution to the formulation of future strategies for tourism in Indian part of the Sundarbans.

TABLE 14.4 Major Problems Encounter for Development of the Sundarbans Tourism.

Nature of problems for tourism development	No. of respondents agreed
Scenery and landscape	10
Fauna and flora	12
Relaxed atmosphere and solitude	7
Climatic conditions	4
Eating facilities	3
Awareness for natural areas	9
Friendliness of hosts	5
Tourist guide availability	6
Quality/availability of accommodation	8
Destination cleanliness	9
Local transportation	5
Safety and security	5
Access to drinking water	6
Availability of tourist information	5
Shopping facility and access	3
Communication infrastructure and access	3
Total	100

Source: Field Survey, 2014.

14.5 DISCUSSION

Due to mismanagement, the natural resources are decreasing. There is a vibrant scope for gentle guidelines for the attention of the very people who depend on the resources. One observation was a surprisingly low awareness of the resource dependency and the resource utilization implications. An improved awareness seems a viable measure, given the fair literacy and education level in the tourism point of view. Another observation was an apparent attractive scope for the introduction of alternative or supplementary livelihoods and income generation, including involvement by the female part of the population. Such development initiatives must—and can—interact with an overruling need of poverty alleviation. Good management and knowledge is compulsory to assure a win–win situation rather than a development where both the mangrove forest and its communities stand to lose. Among other measures in this respect is an improved knowledge base, providing for timely and appropriate decision-making and implementation. This study revealed that due to lack of promotion and advertisement policy, the tourist flow to this destination is not encouraging despite its high potential. This study reveals that there is a lack of uniformity of pricing system of tourism services, causing inconvenience to the tourists when availing the various facilities at the destination. Basic infrastructure facilities to attract tourists like toilets, safe drinking water, etc., are found to be insuffucient in Bishnupur. The study reveals that a complete absence of trained licensed guides at the destination is causing inconvenience to interstate and foreign tourists (Table 14.4).

This study indicates that prices of the land and housing services are astronomically increasing because of the development of tourism in this region. Tourists are found to have no more interest in the quality of basic infrastructure and they expected the development and expansion of tourism-related amenities like tour guides, self-check-in information kiosks, and availability of further sources of reliable information, as it is found to be related with tourist safety measures. Quality/availability of accommodation, basic facility of drinking water, availability of current updated tourist information, and availability of communication infrastructure are found significant with a difference between expectation and perception. There is a need to improve the service providers and the quality of service during service delivery, demonstration, or guiding. It has also been noticed that climatic conditions including high water and low water significantly influence expectation–perception of tourists visiting the Sundarbans.

14.6 RECOMMENDATIONS

This study indicates that tourism education is very much essential to encourage people to engage in tourism entrepreneurial activities in the state. The study depicts that most of the tourism activities in the Sundarbans are mainly indigenous in nature which includes bird watching, fishing, seasonal fairs, festivals, and so forth, which sometimes discourage the tourists to go for repeat visit to the destination. This study reveals that those tourists who are mainly coming from urban areas are trying to exploit the rural culture by degrading the local environment and surrounding. In terms of employment potential, the study reveals that employment opportunity through tourism business is quite low due to the lack of government initiatives, lack of tourism-marketing opportunities, non-availability of trained manpower, and so forth. This study depicts that there is a greater need for accommodation facilities specifically for budget category hotels to accommodate more number of tourists in this destination. A greater need for car-parking facilities at the destination was felt particularly during the peak season. A greater demand for trained tourist guides was felt to provide proper information for the destination. The study reveals that during their visit to the Sundarbans they are more concerned about the safety and security aspects of their tour. Priority should be given to strong law execution to protect waterways and to combat corruption and crime (good governance). Local communities should be encouraged to cooperate in the marketing of their products in order to reduce dependence on traders and "middlemen." Non-government organizations (NGOs) might have a particular role in supporting local communities in finding new sources of income generation. Empowerment of women and supporting their potential role in contributing to family income generation, for example, establishment of small trade and handicrafts, might further increase income security. A potential opportunity lies in developing sustainable tourism alternatives based on community involvement, in close cooperation with park authorities. Education initiatives may additionally increase the scope for a more efficient use of resources.

14.7 CONCLUSION

It is boom time for travel and tourism sector in India. Eastern India has emerged as the frontrunner in its growth chart. West Bengal is replete with business and tourism potential. The travel and tourism sector has entered a take-off stage and it is going to scale a great height in near future. It is growing every day, every month, and every year with a steady pace. But the growth promise must not lead to mindless complacency. At a time when expansion looks assured in the travel and tourism sector, it calls for some hard thinking. Has West Bengal realized its optimum potential in the sector? Has the state government done enough in promoting travel and tourism, particularly tourism in the state? What are its priorities? Has the private sector made adequate contribution to the travel and tourism in the east? Has the center extended a helping hand in this regard?

It does not need any serious analysis to say that despite their immense potential West Bengal do not figure prominently in the tourism map of India. Internationally the Eastern India is yet to make a true mark. The number of domestic and foreign tourists visiting West Bengal and the neighboring states has been on the rise in the last few years. The rising trend may yield some satisfaction. But a comparison with other states and regions in India shows that the east region, particularly West Bengal, have failed to measure up to their potential. The reason is simple. While other states are doing all around publicity to attract tourists, West Bengal is lagging behind in its visibility campaign.

KEYWORDS

- **the Sundarbans**
- **mangrove**
- **expectation**
- **perception**
- **Wilcoxon test**

REFERENCES

Chaudhary, M. India's Image as a Tourist Destination- a Perspective of Foreign Tourists. *Tour. Manag.* **2000,** *21* (3), 293–297.

Kamra, K. K. *Managing Tourist Destination: Development, Planning, Marketing, Policies;* Kanishka, Publishers & Distributors: New Delhi, India, 2001; p 1–47.

Litvin, S. W.; Ling, S. N. S. The Destination Attribute Management Model: An Empirical Application to Binton, Indonesia. *Tour. Manag. London.* **2001,** *3* (5), 481–492.

Moscardo, G. Community Capacity Building: An Emerging Challenge for Tourism Development. In *Building Community Capacity for Tourism Development;* Moscardo, G., Ed.; CABI: UK, 2008.

Nicolau, J. L.; Mas, F. J. Micro Segmentation by Individual Tastes on Attributes of Tourist Destinations. In *Tourism Management: New Research New York;* Liu, T. V., Ed.; Nova Science Publishers: New York, NY, 2006.

Pearce, P. L. *Tourist Behavior: Themes and Conceptual Schemes;* Viva: New Delhi, India, 2000.

Rahman Mohammed, M.; Rahman, M. M.; Islam Kazi, S. The Causes of Deterioration of Sundarban Mangrove Forest Ecosystem of Bangladesh: Conservation and Sustainable Management Issues. *AACL Bioflux*. **2010,** *3* (1), 77–90.

CHAPTER 15

WORLD AROUND WILDLIFE TOURISM

JOHRA K. FATIMA[1*] and MAHMOOD A. KHAN[2]

[1]*School of Management, University of Canberra, Canberra, Australia*

[2]*Virginia Polytechnic and State University, Blacksburg, VA, USA*

[*]*Corresponding author. E-mail: johra.fatima@canberra.edu.au*

CONTENTS

ABSTRACT

Wildlife tourism research deals with a variety of issues including attitude measurement of local community, visitor support system and management, government policies, economic and ethical perspectives, and so forth. However, researchers on this tourism stream need to re-conceptualize the current understanding and scope of wildlife tourism by configuring its greater impact on environment, economy, society, and sustainability. Considering this overwhelming impact of wildlife tourism, more consolidated research from multiple disciplines with multi-country efforts are highly encouraged.

> **"There are still many undiscovered trails in the deep jungle of wildlife tourism….."**

In wildlife tourism, local community plays a vital role for ensuring long-term success and proper manintainance of tourism destinations (Chandralal, 2010). Without positive attitude and support of local community, often it is impossible to implement wildlife tourism policies, visitor management, and protect bio-diversity (Schofield et al., 2015; Gössling, 1999; Allendorf, 2007). In such case, awareness and engagement of the community with the tourism activities becomes a fundamental issue of concern. Wildlife recreation generates a major economic source of income for local community as well (Goodwin, 2002; Zografos & Oglethorpe, 2004). Therefore, development of local infrastructure, facilities and other direct financial benefits will help to acquire more local attention, and will enhance community engagement. As it is necessary to establish a strong interactive and productive relationship among wildlife tourism activities and local community, several approaches need to be considered for ensuring this including developing, and implementing realistic policies (Sirakaya, 1997; Duprey et al., 2008; Schofield et al., 2015; Donnelly et al., 2011). However, if any aggressive or unenthusiastic phenomenon arises from local community for hampering wildlife, it can be resolved by alerting them about the fragile environment, interdependence between wildlife and community, effect of human activities on wildlife populations, and so forth. (Hughes et al., 2011; Turley, 1999; Mason, 2000).

There are also many less attended and overlooked tourism destinations, which could be nourished to turn these in attractive tourism destinations.

In particular, it is difficult for developing countries to take appropriate initiatives for lack of investments, poor marketing efforts, and inadequate infrastructure facilities. Government has to take leading role to provide a primary platform for the private entrepreneurs to take initiative for the development of popular tourism destinations. Investments on infrastructure development, considering measure for safety, and security issues, latest technological assistance for visitor information, interactive communication, and image building may help to attract visitors to these hidden treasures of a country. If a single tourism destination has been located in more than one country (such as the Sundarbans in Bangladesh and India), it would be useful to undertake combined policies, collaborative research, strategic partnership, and marketing initiatives by all related governments. This will reduce cost and efforts as well as will make stronger impact on global tourists mind. In such kind of marketing initiatives, targeted market research is necessary to explore specific travel motives, factors influencing destination branding and identifying appropriate positioning strategy. Research should also be conducted on the meaning of place attachment and sense of place of a specific tourism destination for the targeted visitors (Amsden et al., 2011). While setting marketing objectives, emphasis should be given on cultural, history, or natural resources of the destinations (López-Guzmán et al., 2011) along with a differentiate strategy which will offer a complete new experience for the visitors.

While it is crucial to focus on unexplored and ignored travel locations for making these as attractive tourist destinations, emphasis should also be given to the distinctive non-consumptive wildlife tourism activities. For instance, bird watching and marine life tourism activities are gradually creating huge curiosity among wildlife visitors. Scholars (Newsome et al., 2002; Bruce & Bradford, 2013; Vianna et al., 2012; Burgin & Hardiman, 2015) have confirmed the increased participation rate for marine wildlife. Marine wildlife tourism mostly dominated by the watching of whales, dolphins, penguins, marine turtles, and sharks (Schofield et al., 2015; Anderson et al., 2011). Visitor participation for viewing or swimming with cetaceans has been grown around 13 million worldwide, which is 30.5% compared to the figure in 1998 (O'Connor et al., 2009; Burgin & Hardiman, 2015). Similarly bird watching becomes a popular globally activity (Maple et al., 2010) as well. Visitors get more excited if bird watching is accompanied by offering opportunity of watching scenic beauty, variety of birds, hiking, and other recreational amenities (Scott &Thigpen, 2003).

However, there are many unethical and disturbed behaviors often conducted by visitors, local community, and other third parties, which put serious thread for the existence of wildlife. In case of marine wildlife, marine litter particularly, plastic bags and bottles, metals, fishing gears, cigarette butts make threads of living for the marine lives, and sustainable marine environment (Wilcox et al., 2016; Page et al., 2004). Killing and trade of wildlife is another thread for long-term viability of the population of wildlife especially for the endanger species (CPW, 2014). To lessen and to remove the human disturbance and unethical behavior, it is necessary to emphasize on public awareness and issues related with animal ethics (Hughes et al., 2011). Explaining the interrelationships of wildlife-habitats and describing the devastating effect of human unethical behavior on the long-term sustainability of wildlife and bio-diversity may be effective ways to increase public awareness (Mason, 2000; Turley, 1999).

In conclusion, there are still many undiscovered trails in the deep jungle of wildlife tourism, which need to be explored. The beauty of wildlife tourism is that it always incorporates different branches of knowledge stream to reach to a consolidated conclusion. This interdisciplinary nature of wildlife tourism research allows scholars from different disciplines to work and share their precious experiences with one another in a common platform.

KEYWORDS

- **wildlife tourism**
- **environment**
- **economy**
- **society**
- **sustainability**

REFERENCES

Allendorf, T. D. Residents' Attitudes toward Three Protected Areas in Southwestern Nepal. *Biodivers. Conserv.* **2007,** *16,* 2087–2102.

Amsden, B. L.; Stedman, R. C.; Kruger, L. E. The Creation and Maintenance of Sense of Place in a Tourism-Dependent Community, *Leisure Sci.* **2011,** *33,* 32–51.

Anderson, R. C.; Shiham, A. M.; Kitchen-Wheeler, A. M.; Stevens, G. Extent and Economic Value of Manta Ray Watching in Maldives. *Tour. Mar. Environ.* **2011,** *7,* 15–27.

Bruce, B. D.; Bradford, R. W. The Effects of Shark Cage-Diving Operations on the Behaviour and Movements of White Sharks, Carcharodon Carcharias, at the Nepture Islands, South Australia. *Mar. Biol.* **2013,** *160,* 889–907.

Burgin, S.; Hardiman, N. Effects of Non-Consumptive Wildlife-Oriented Tourism on Marine Species and Prospects for their Sustainable Management, *J. Environ. Manage.* **2015,** *151,* 210–220.

Chandralal, K. P. L. Impacts of Tourism and Community Attitude towards Tourism: A Case Study in Sri Lanka. *SAJTH.* **2010,** *3*(2), 41–49.

CPW 2014; Sustainable Wildlife Management and Bio-diversity, Retrieved on, 4.3.2016, http://worldparkscongress.org/drupal/sites/default/files/documents/docs/CPW_FS_Biodiversity_web.pdf

Donnelly, R. E.; Katzner, T.; Gordon, I. J.; Gompper, M. E.; Redpath, S.; Garner, T. W. J.; Altwegg, R.; Reed, D. H.; Acevedo-Whitehouse, K.; Pettorelli, N. Putting the Eco Back in Ecotourism. *Animal Conserv.* **2011,** *14,* 325–327.

Duprey, N. M. T.; Weir, J. S.; Wursig, B. Effective-ness of a Voluntary Code of Conduct in Reducing Vessel Traffic Around Dolphins. *Ocean Coast. Manage.* **2008,** *51,* 632– 637.

Goodwin, H. Local Community Involvement in Tourism around National Parks: Opportunities and Con- Straints. *Curr. Issues Tour.* **2002,** *5,* 338–360.

Gössling, S. Ecotourism: A Means to Safeguard Bio- Diversity and Ecosystem Functions? *Ecol. Econ.* **1999,** *29,* 303–320.

Hughes, K.; Packer, J.; Ballantyne, R. Using Post-Visit Action Resources to Support Family Conservation Learning Following a Wildlife Tourism Experience. *Environ. Edu. Res.* **2011,** *17* (3), 307–328.

López-Guzmán, T.; Sánchez-Cañizares, S.; Pavón, V. Community - Based Tourism in Developing Countries: A Case Study. *Tourismos: Int. Multidiscipl. J. Tour.* **2011,** *6* (1), 69–84.

Mason, P. Zoo Tourism: The Need for More Research. *J. Sustain. Tour.* **2000,** *8* (4), 333–339.

Maple, L. C.; Paul, F. J. E.; Heather, R. Birdwatchers' Specialisation Characteristics and National Park Tourism Planning. *J. Ecotour.* **2010,** *9* (3), 219–238.

Newsome, D., Dowling, R.; Moore, S. *Wildlife Tourism;* Channel View Publications: Clevedon, UK, 2005.

O'Connor, S.; Campbell, R.; Cortez, H.; Knowles, T. Whale Watching World- Wide: Tourism Numbers, Expenditures and Expanding Economic Benefits. International Fund for Animal Welfare: Yarmouth, MA, USA, 2009.

Page, B.; McKenzie, J.; McIntosh, R.; Baylis, A.; Morrissey, A.; Calvert, N.; Goldsworthy, S. D. Entanglement of Australian Sea Lions and New Zealand Fur Seals in Lost Fishing Gear and other Marine Debris before and after Government and Industry Attempts to Reduce the Problem. *Mar. Pollut. Bull.* **2004,** *49,* 33–42.

Schofield, G.; Scott, R.; Katselidis, K. A.; Mazaris, A. D.; Hays, G. C. Quantifying Wildlife-Watching Ecotourism Intensity on an Endangered Marine Vertebrate. *Animal Conserv.* **2015,** *18,* 517–528.

Scott, D.; Jack, T. Understanding the Birder as Tourist: Segmenting Visitors to the Texas Hummer/Bird Celebration. *Hum. Dimens. Wildl.* **2003**, *8* (3), 199–218.

Sirakaya, E. Attitudinal Compliance with Ecotourism Guidelines. *Ann. Tour. Res.* **1997,** *24,* 919–950.

Turley, S. K. Conservation and Tourism in the Traditional UK Zoo. *J. Tour. Stud.* **1999,** *10* (2), 2–13.

Vianna, G. M. S.; Meekan, M. G.; Pannell, D. J.; Marsh, S. P.; Meeuwig, J. J. Socio-Economic Value and Community Benefits from Shark-Diving Tourism in Palau: A Sustainable Use of Reef Shark Populations. *Biol. Conserv.* **2012,** *145,* 267–277.

Wilcox, C.; Nicholas J.; Mallos, G. H.; Leonard, A. R.; Hardesty, B. Using Expert Elicitation to Estimate the Impacts of Plastic Pollution on Marine Wildlife. *Mar. Policy.* **2016,** *65,* 107–114.

Zografos, C.; Oglethorpe, D. Multi-Criteria Analysis in Ecotourism: Using Goal Programming to Explore Sustainable Solutions. *Curr. Issues Tour.* **2004,** *7,* 20–43.

INDEX

K

L

M

N

O

P

T

U

V

W

For Product Safety Concerns and Information please contact our EU representative GPSR@taylorandfrancis.com Taylor & Francis Verlag GmbH, Kaufingerstraße 24, 80331 München, Germany

T - #0153 - 230425 - C310 - 229/152/14 - PB - 9781774636916 - Gloss Lamination